Decades of Decadence

ALSO BY MARCO RUBIO

American Dreams: Restoring Economic Opportunity for Everyone

An American Son: A Memoir

Decades of Decadence

How Our Spoiled Elites Blew America's Inheritance of Liberty, Security, and Prosperity

MARCO RUBIO

BROADSIDE BOOKS

FIRST EDITION

Library of Congress Cataloging-in-Publication Data has been applied for.

ISBN 978-0-06-329697-8

23 24 25 26 27 LBC 5 4 3 2 1

To my parents, who came to America and lived the American Dream, and to my children and their generation, for whom I fight every day to make sure we don't lose it.

It was scarcely possible that the eyes of contemporaries should discover in the public felicity the latent causes of decay and corruption. . . . Their personal valor remained, but they no longer possessed that public courage which is nourished by the love of independence, the sense of national honor, the presence of danger, and the habit of command. . . . The posterity of their boldest leaders was contented with the rank of citizens and subjects.

—*Edward Gibbon,* The History of the Decline and Fall of the Roman Empire

Contents

Introduction: The Return of History xi

Chapter 1 A Failed Consensus 1

Chapter 2 Paper Wealth 23

Chapter 3 The Rise of China 48

Chapter 4 Crisis of Confidence 67

Chapter 5 The Rise of the Experts 88

Chapter 6 American Nationalism 110

Chapter 7 Anti-America, Inc. 135

Chapter 8 The Liberal Culture War 156

Chapter 9 American Marxism in Action 176

Chapter 10 A Time for Choosing 193

Acknowledgments 211

Notes 213

Index 215

Introduction: The Return of History

In the summer of 1989, just as I was getting ready to go to college, a small foreign policy journal called the *National Interest* published its sixteenth issue. Normally, this wouldn't have caused much of a stir. It certainly isn't the kind of event that you'd expect to see on the first page of a book written more than thirty years later.

But this particular issue was different. Inside, a deputy director at the State Department named Francis Fukuyama published a long article titled "The End of History?" Over the course of about sixteen pages, the author expounded on world history, political philosophy, and his ideas about the future of global order. The language was dense, and the ideas, many of them borrowed from the German philosopher Georg Wilhelm Friedrich Hegel, weren't exactly the kind of thing you'd be excited to hear someone bring up at a dinner party.

The title was catchy. And at the time, so was the main argument, which said, in short, that after centuries of conflict among incompatible political and economic systems, liberal democracy—i.e., the government of the United States and our European allies—had won. There were no alternatives to democracy left standing. The United States had defeated fascism during World War II, and now, as the Soviet Union was on the verge of collapse, it seemed that we were about to experience the end of communism, too.

"What we may be witnessing," according to Fukuyama, "is not just the end of the Cold War, or the passing of a particular period of postwar history, but the end of history as such: that is, the end point of mankind's ideological evolution and the universalization of Western liberal democracy as the final form of human government."

Given global events, these words seemed prescient. A few months earlier, in December of 1988, Mikhail Gorbachev had gone before the United Nations General Assembly to announce that the Soviet Union would soon cut funding for its military by almost 15 percent. Thousands of Russian troops that had once been stationed in Soviet satellite states would soon march home, leaving those countries free to adopt the principles of democracy. During this address, which included sober quotations from the English poet John Donne, Gorbachev declared that "the use of threat of force no longer can or must be an instrument of foreign policy."

The Cold War, in other words, was over. The United States had won. America's elites could sit back, relax, and enjoy the spoils of victory. Concepts like nation-state competition and realities of the human condition that had proven true for 5,500 years of recorded history, such as the struggle with the desire of the powerful to conquer the weak, could be relegated to "the museum of human history," where they would be studied and contemplated since they were no longer relevant to governing modern societies.

It would not take long for this idea to spread not only to Washington think tanks and college seminar rooms but to the very highest levels of the US government. Within a few years, the sense that history was over came to change the way US policymakers thought about our place in the world. Rather than working to assure that the United States would maintain its internal strength and its position as the world's dominant superpower, our leaders enacted policies that put this country on a road of slow, inevitable decline.

Rather than shoring up our manufacturing capacity, businesses exported jobs to places like Mexico and China, leaving many American workers without the means to provide for their families. Rather than placing our trust in institutions, we took them for granted and began carelessly destroying those institutions. And rather than enacting policies that put key values such as family and community at their center, we came to focus on consumption above all else. Americans, who see themselves as workers, fathers, and citizens, came to be viewed by policymakers as simply consumers who bought products—nothing more.

This is a book about what happens when the country's elite decide that all the things that brought our nation this success are, in fact, the things we most urgently need to bring to an end. The "end of history" began a conflict in American society that we still live through every day, a conflict we consistently get wrong. We fail to see the big picture. We know the elite say it doesn't matter if some towns lose all the jobs and collapse. In fact, this is necessary. The elites say we don't need traditional families anymore. The elites say you can find everything you need online. People with good jobs and strong families building their local communities are, they tell us, not America's greatest resource, but an outmoded way of living that needs to be left behind. We are all citizens of the world, and America's leadership is no longer needed or wanted. The qualities that gave this country strength for over two centuries, they tell us, are now the sources of great weakness.

For most of us, we did not realize in the 1990s that the next battle would be fought on these grounds. In the pages that follow, I'm going to argue that we continue to lose this battle because we still fail to see the way it all ties together. They are attacking on four fronts: good local jobs, stable families, geographical communities, and a nation that serves as a beacon of freedom and prosperity. When they weaken one of these, they weaken all of these. That's the point.

How can the elite have gotten it so wrong? They believed that as long as we could make other countries—specifically China—adopt the principles of capitalism, liberal democracy, and progressive values, then those countries would soon come to look just like the United States. Neoliberalism, in this sense, meant a belief in free trade above all else. It meant wide-open borders, which would ease the flow of goods and people. It meant cheering on politicians and candidates who made these their causes. It meant attacking and rolling back any values seen as retrograde, like a belief in equality of opportunity, not outcomes, or the existence of two—and only two—genders. This ideology was hostile to even the slightest whiff of patriotism; it downplayed or outright denied the greatness of our nation and the existence of American exceptionalism. Once all the globe had submitted to this elite ideology, our leaders believed that the world would become a liberal democratic utopia, and the only remaining work would be deciding how to spend our days.

At home, this led to a sustained period of decadence—a term characterized by conservative *New York Times* columnist Ross Douthat as "economic stagnation, institutional decay, and cultural and intellectual exhaustion at a high level of material prosperity and technological development." During this period, the cost of some luxuries like flat-screen TVs has dropped, but critical goods like homes and health care have become prohibitively expensive for millions of Americans. At the same time, technological innovation outside of the internet—in other words, in the real world—has dropped, and our culture and politics have become listless. The result is unprecedented levels of apathy and unpatriotic feeling, especially among young people. Our decadent elites have come to believe that things that used to matter—human nature, patriotism, and community—don't matter anymore. They believe the era of global conflict is over, and the only job left to do is sit around, get rich, and endlessly criticize our history.

Over the course of this book, you will hear stories you may be familiar with but in a new and different light. We'll look at the nefarious actions of the Chinese Communist Party (CCP), the decline in academic freedom, and the never-ending push for open borders. Two key dynamics will emerge. The first is that the external threats to our way of life that we survived during the Cold War are still very much with us. Thanks to my upbringing, I knew that better than most of my peers.

In Cuba, for instance, the country my grandfather had fled in 1956 for the shores of the United States, Fidel Castro was growing even more committed to the socialist, anti-Western cause than ever. He was preparing his people for a long, drawn-out period of austerity, after which, he claimed, they would return to prosperity. And the people in his regime, much to the surprise of Western observers, were willing to follow him through years of hell to get there.

According to the foreign policy elites in Washington, this shouldn't have happened. In their minds, the struggle of the Western world against tyranny and socialism was over. Communist governments were crumbling all over the world, and they had no reason to believe that the trend wouldn't continue. According to their position papers and foreign policy journals, the collapse of the Soviet Union—which until very recently had supplied Cuba with almost all of its oil, fertilizer, and grain, as well as most of the island's military protection—should have brought the Castro regime tumbling down soon afterward.

A conservative group called the Cuban American National Foundation went so far as to write a new constitution for the soon-to-be-liberated post-Castro Cuba. They assumed the Cuban people would need one any minute, given that Castro was, in the words of a popular author at the time, a "dead man walking." A few months later Congress passed the Cuban Democracy Act, which tightened restrictions on Cuba to a degree that would, in the words of Congressman Bob Torricelli, bring Castro "to his knees."

Those policies were long overdue, but they did not bring an end to the regime. Instead, Castro only became further entrenched, and he expected his people to follow his lead. As a proud man who led a country of people who were proud of their heritage, he was not willing to give up the fight. Evidently, no one had bothered to ship him the *National Interest* issue declaring history to be over. In the early 1990s, with his country suffering from inflation, rolling blackouts, and rising crime rates, Castro began ending all of his speeches with the phrase "Socialism or death!"

For him, history was just beginning.

And it wasn't just Cuba.

Around the same time in East Germany, a young KGB agent named Vladimir Putin was called home from his post in the city of Dresden, where he'd been stationed for about six years. The Soviet Union, he learned, had just voted itself out of existence, and his services as a soldier and spy would no longer be required. According to the consensus in Washington, DC, this young man should have rented an apartment in Moscow, gotten a job driving a taxi, and awaited the inevitable advent of liberal democracy. The strong impulse toward authoritarianism that had haunted Russia for centuries, in the eyes of the Washington elite, would no longer be a problem. Russia, in time, would become part of the global market. There was even talk of having them join NATO.

But that's not what happened. While he did spend a few years driving a taxi around the streets of Moscow, Vladimir Putin did not accept the fall of the empire he had served for sixteen years with the quiet shrug expected of men like him. He didn't move on and submit to a liberal democratic government. Instead, he fumed about the collapse of his homeland's government, vowing in silence to return the Soviet empire to its former glory. In years to come, he would refer to the fall of the Soviet Union as "the greatest geopolitical catastrophe of the century."

Similarly, in China, where the government had more success in attracting and coercing American and multinational companies into its market, the Communist Party was growing more repressive than ever. According to the experts in Washington, if we could only get the country to adopt the principles of Western capitalism, helping them to become rich in the process, it would be only a matter of time before the government threw off its authoritarian tendencies and adopted the principles of liberal democracy too. For the past decade presidents and their representatives had approached China in this way, almost as if they had already become the rational, liberal democracy we imagined them to be.

How could anyone truly believe history was over following the events of the spring of 1989—just months earlier—when thousands of pro-Western demonstrators gathered in Tiananmen Square in Beijing? For months, they protested the repressive Communist regime that was in control of their country. In response, the Chinese government declared martial law and sent in the military. Over the course of about two days in early June, soldiers shot protestors with assault rifles and ran them down with tanks. It is estimated that anywhere between four hundred and three thousand people died during the massacre. Of course, given the total media blackout in China after the incident—something that occurs only in authoritarian countries—we'll never be sure of the real number.

As the world looked on in horror, the Chinese leadership took no steps to modernize its government, or to implement Western-style democracy. Even as China became more integrated with the global economy—largely thanks to the help of the United States—its government became yet more oppressive, finding clever new ways to hide information from its citizens and exert greater control over their lives. Clearly its leaders' plans did not include submitting to the end of history and becoming a Western-style democracy. What they *did* include was world domination.

It's worth pausing here to go back to and reflect on Putin's words. No amount of wealth was going to change his worldview. No investment account was going to make him care about others. No membership in systems of global finance was going to make him see Russia as a team player in America's new world order. To Vladimir Putin, the fall of the Soviet Union was not a disaster because of some material concern he had. It was a disaster because place matters. History, culture, and citizenship all have value that is more intrinsic than economic concerns.

For much of the past thirty years, we have assumed that money can change the world for the better. We have assumed that material wealth will make people like Putin embrace our values. And along the way, we've forgotten that it is our values that have made America great and prosperous, not the other way around. The "prosperity gospel" preached by the likes of Fukuyama is a direct rejection of

America's founding. It is a direct rejection of fundamental truths. It is also incredibly naive and dangerous.

To people living in China and Russia—not to mention North Korea, Venezuela, or a whole host of other nations—history did not end with the fall of the Berlin Wall. The collapse of a few Communist regimes in the early 1990s did not, in their eyes, mean that it was time to accept a new form of government and abandon their own cultures, no matter how many position papers or policy articles said that this was happening.

These were cultures with rich historical traditions that were not willing to submit to a "flattening of the world," as Thomas Friedman put it in the *New York Times*. Some countries were not ready to accept free trade capitalism, liberal democracy, or progressive values. In others, such a system seemed antithetical to their history and beliefs.

Some countries, as we would soon learn, were willing to fight to make sure history was not over.

For a while, the United States could afford to pretend that this was not the case. For a few years we could enjoy what the writer Charles Krauthammer called the "unipolar moment," where, in his words, "world power resides in one reasonably coherent, serenely dominant, entity: the Western alliance, unchallenged and not yet (though soon to be) fractured by victory."

Again, at the time, we were living in a brief unipolar moment. When Saddam Hussein invaded Iraq's neighbor, Kuwait, on August 2, 1990, the United States, leading an alliance of nations, succeeded in pushing his military out of that nation in a mere one hundred hours. Saddam's army at the time, battle-hardened after a decade of war with Iran, was considered one of the top armies in the world. The Gulf War remains one of the most stellar military victories that the world has ever seen. In the afterglow of such a victory, we believed that the dominance of the Western alliance was inevitable. Our military was unmatched, and the horrors of war were things of the past. Now, war was something you watched on CNN.

The idea that a repressive regime like China or Russia could pose any major threat to a global order run by the United States was foreign to us—certainly to the polite, well-educated elites who were making decisions about US foreign policy at the time, and continue to do so to this day. Over time, these people have gone from representatives of the most powerful country in the world to the meek, apologetic caretakers of America's decline. The problem is not the recognition that we have experienced a unipolar moment. The problem is what we have chosen to do with it.

Which brings us to our second, deeper issue. Is our internal collapse a greater threat? Because of decadent choices we have made for thirty years, today we are weaker, more fractured, and less confident than we ought to be. It started when the elites of both parties began to tell us that better jobs and cheaper goods would make up for the jobs we'd lost to Mexico and China. Now, instead of providing those good local jobs, they've told us we need to move. Instead of encouraging stable two-parent families, they've told us that any people who love each other can form a family. Instead of encouraging participation in our communities, they've told us that online activism is enough. Instead of saying that America is a beacon of liberty and prosperity, they've told us that we're better off as citizens of the world rather than of this oppressive nation.

There is a common theme here: a globalist elite hell-bent on undermining America's best traditions and weakening a nation it views as an overly religious, backward, and imperialist regime unwilling to embrace a new world order. But now is not the time for weakness. As I write these words, the armies of Vladimir Putin have invaded Ukraine, a country that Putin believes is vital to restoring the former glory of the Soviet Union, because he assumed Ukraine would fold and the West would stand by and watch. China is engaging in genocide, and its representatives are beginning to carry out long-held plans to supplant the United States as the world's preeminent superpower. Iran, which has spent the past decade striving to build an arc of dominance stretching from Lebanon through Syria and

Iraq, continues to make strides in developing nuclear weapons that it believes will allow it to act with impunity on the world stage.

Meanwhile, the United States has grown complacent. After years of globalization, our towns have been hollowed out by the loss of factories and jobs. It is no longer possible for the vast majority of Americans to raise a family on a single earner's wages. Our military, which has spent two decades mired in conflicts in the Middle East, is failing to meet its recruitment quotas. In towns all over the United States, an opioid epidemic is tearing families and communities apart. During the Fourth of July weekend in 2022, nine people died in a single county in Florida after ingesting small amounts of fentanyl—a drug that is largely manufactured in China and imported into the United States illegally through our open southern border.

The response from our country's elites—the same ones who have created these problems—has been mostly silence. In the White House, President Joe Biden is unwilling to address the major problems of our time. Instead, he has pursued the most inflationary economic policies in modern history, driving the costs of consumer goods to record levels. He has given in to the demands of the radical left wing of his party on climate and green energy, sending a message to fossil fuel companies that drilling for more oil and natural gas is not worth their time. Thanks to his incompetence, gas prices fluctuated wildly during his first years in office.

Rather than using this dire moment to bring our nation together, our elites are focused on shoving far-left culture war issues down Americans' throats and self-flagellating over our history. For the past decade we have torn down statues of our founding fathers and talked endlessly about the negative aspects of their legacies. We have begun teaching children that the United States is nothing more than a nation of pirates built on the principle of white supremacy. Anyone who dares to disagree runs the risk of being called a racist or a bigot in public, a charge that has become nearly impossible to deny or disprove.

Sadly, the world has noticed.

During an address to the United Nations General Assembly in January 2021, Linda Thomas-Greenfield, who'd just been sworn in as President Biden's ambassador to the United Nations, made sure to mention that the United States of America had an "original sin," which was slavery. She also spoke of "the senseless killing of George Floyd," "white supremacy," and the "spike in hate crimes over the past three years."

So it should come as no surprise that two months later, when Secretary of State Anthony Blinken met for the first time with his Chinese counterparts, he was forced to hear about the "many problems within the United States regarding human rights . . . such as Black Lives Matter." As a result of these moral evils—which the United States admitted to on its own—this representative of the Chinese government said that the United States "does not have the qualification to say that it wants to speak to China from a position of strength."

What's troubling about this statement is not the mere fact that the Chinese government—one of the most repressive, abusive regimes in the world—was allowed to embarrass the United States on the world stage, although that is certainly a problem. It is that the elites governing our nation and its most powerful institutions actually agree with the Chinese propaganda. Most Americans know this is grotesque nonsense, but to the small minority that run many of our most important companies, shape our media dialogue, and write or implement the laws of our nation, there is a fervent belief that the United States is an evil nation, and that there is nothing good about our country worth fighting for.

This is an opinion that can only be formed in the most decadent society over a period of relative peace, which is exactly what we've had for the past three decades.

That era is now over. It is not over because we have decided to stop this behavior, but because the global order has again changed. Russia has declared its intention to assert a sphere of influence, and

China seems poised to help them do so. Meanwhile, the government in Beijing has begun spreading its Belt and Road Initiative—a form of economic imperialism that should be deeply troubling to those in the West—to countries all over the world.

It is not too late to free ourselves from the errors of three decades of decadence and confront these challenges of our day. Despite the events of the past thirty years, America remains a powerful and wealthy nation built on timeless ideals ingrained in the very creation of mankind.

But to seize on our advantages we must act decisively. We cannot get things as wrong in the years ahead as we have over the past several decades. Indeed, it is time for new leadership, because for the past thirty years, the polite and orderly caretakers of America's decline have had their moment. They failed to leave America stronger than what they inherited. They saw China's rise, but they didn't see a problem with it. They didn't prepare a nation for hard times. They plundered our economy and sent it off to the rest of the world. They scorned traditional American values, prizing short-term consumption and identity politics over intact families, strong communities, and national pride. They sought to replace democracy at home with a technocracy of experts, and in doing so tremendously diminished our nation's faith in its institutions. And they have been incapable of articulating a focused foreign policy that is prepared to deal with the economic and national security threats our nation faces.

We must face the return of history guided by leaders who understand the threats and the inherent evil behind those threats. It is something I am reminded of every day in Miami. In certain neighborhoods you cannot get a cup of coffee without running into someone who knows that evil firsthand. They know the costs of apathy and authoritarianism. America and the free world cannot afford another misguided and decadent decade.

This book lays out the critical moments of the last thirty years. Across its chapters, I aim to identify and elaborate on the three core failings of progressive liberalism today: (1) its opposition to and destruction of timeless American institutions and parts of the economy for short-term gain; (2) its fundamental misunderstanding of the rest of the world, including other nations' enthusiasm for suddenly adopting radically new economic and social values; and (3) its arrogant confidence in its own value system, which has led to countless destructive culture wars at home. Only by properly understanding these errors can we learn from our history and embark on a better way forward. For in the years to come, we will either enter a new dark age, or we will usher in a New American Century.

Decades of Decadence

Chapter 1

A FAILED CONSENSUS

As with most people who have lost a parent, not a single day goes by that I don't think about mine. I lost my father in 2010, two months before I was elected to the US Senate. My mother passed away in the fall of 2019. Most days, it is some small interaction I have with my family that reminds me of my parents—a lesson I try to teach one of my children; a parenting challenge Jeanette and I face together; something I use to see my parents do that at the time seemed silly to me, but now makes me laugh as I do the exact same thing in my own household.

But other times, I think of my parents when I am in a very different context than my homelife in Miami. When I am in Washington, DC, sitting in my office in the US Capitol complex, I sometimes reflect on the lives my parents lived and the America they came to. At these moments, I am often struck by how different their experience of the American Dream was from what it is today.

Many Americans who follow politics closely have heard my life story. I am the son of two hardworking immigrants from Cuba. My father worked primarily as a bartender. My mother stayed at home when we were young, but spent time working in hotels as a maid and briefly in a local factory. My parents worked hard and lived responsibly so that they could own a home; provide their children the security, stability, and opportunity to achieve our dreams; and eventually retire with dignity.

My parents lived the American Dream. It wasn't about get-rich-quick schemes, luxury cars, or fame. It was about family, faith, and community. That is the backbone of a strong nation. I ran for president of the United States in 2016 with a fervent desire to fight for this dream. Yet what became increasingly clear to me over the course of that campaign is that, for way too many Americans, that dream is not only out of reach but nearly impossible even to imagine. This country has undergone immense economic and social changes since my parents first came to this country. Many of these changes have not been for the better.

Today, a married couple working honest working-class jobs as a bartender and a maid would not have the financial security to own their own home, or allow a mom to stay at home with the kids. Not only does the data show that—but everybody knows it. Frankly, most people wouldn't expect that a couple working those jobs would be married in the first place, let alone be able to afford raising four kids in a stable environment, or send their kids to a good school, like my parents did for me.

Today, too many hardworking Americans who want to do the right thing and live normal, decent lives are unable to do so. And Americans know it. It is not nostalgia to crave a home, family, and stability. But that dream—one on which the middle class in this country was built—is getting harder and harder to achieve.

Obviously, this decline in the earning potential of American families can be traced to many bad decisions over the past thirty years or so. But I believe the most important moment came in 2001, when the United States helped to ease the Chinese Communist Party's entry into the World Trade Organization (WTO)—the group that sets the standards for trade among nations. This helped China to become a major economic player on the world stage, and it supercharged the decline of American manufacturing. The decision to do this—a decision that was made over a period of many years by elected officials and "experts" of both parties—stands as the clearest example of the elite political class embracing the new doctrine of neoliberalism without properly taking into account the consequences that their actions would have.

To emphasize just how far we have fallen as a result of this decision, I would like to return for a moment to my own family.

Labels

For a period of time when I was young, my mother found work at a factory in a city in South Florida called Hialeah. The factory produced metal folding chairs. Every morning she would go to work with hundreds of men and women in the area, clocking in at nine o'clock and returning home on the bus shortly after five.

At the time, jobs like these were enough to support a family in one of the many small suburban neighborhoods nearby. People could work for forty years, earn a decent wage, and retire with dignity at the age of sixty-five with a pension. Fortunately for us, thanks to my father's long hours tending bar, my mother didn't have to support our whole family with her income, so she could afford to cut her hours when circumstances called for it.

For us and for our neighbors, nearly all of whom were recent Cuban immigrants, these jobs were a godsend. It meant that they didn't have to rely on government assistance to supplement their paychecks. It meant they could come home at the end of the day and look at their children with the satisfaction that comes from knowing that they would not go hungry or work two jobs just to make ends meet.

But beyond financial security, there was a sense of pride in the work these people did. Although the jobs themselves were often repetitive and rarely glamorous—many of them involved putting the same two or three pieces together all day on the concrete floor of a factory, with a few minutes for lunch in between—they were necessary. And they were noticeable. The fruits of the workers' labor were visible in stores and households all over the East Coast. If you turned over a metal folding chair or a plate in many places in Florida, for instance, you'd see a white label with small block letters on the bottom: *Made in Hialeah*.

When I was growing up, I didn't think much about those labels. I'm sure I didn't even notice that much of the cutlery, furniture, and kitchenware in our ground-floor apartment was manufactured just a few miles down the road at the Hialeah factory. When I went to a backyard barbecue at a relative's house, I probably didn't stop to wonder whether my mother—or one of my neighbors, or another one of my relatives—had played some part in making all the items I saw around the house. Back then, we took it for granted that most of the things we bought and used were made right here at home by our fellow Americans.

But we shouldn't have.

Today, the job my mother did at the Hialeah factory no longer exists. Most of the few remaining jobs like it don't pay nearly enough for one person to support a family. In 1985, according to the "Cost of Thriving Index" produced by Oren Cass, it took the median male worker in the United States precisely forty weeks to earn enough money to afford a year's worth of the basics that it takes to raise a family. Today, that same index indicates it would take sixty-two weeks in a fifty-two-week year to do the same thing. Think about that: there aren't enough weeks in the year for a normal American worker to provide for life's basic necessities. Yet according to America's foremost economic experts, America is wealthier than ever. "Listen to us, don't believe your lying eyes," they tell the rest of us.

Today, if you go to a backyard barbecue in Miami and start turning over plates, towels, and pool toys, it's unlikely that you'll see the *Made in Hialeah* label on any of them. Instead, you're almost guaranteed to see *Made in China*. In fact, at least half the items in any given room in the United States probably carry the same label. While it was once the case that most of what we consumed in the United States was made right here at home—often just a few miles away at the very mills and factories that sustained our local communities—that's no longer true.

Why?

It stems from a decision that the elites of this country made about twenty years ago to open the American economy to Communist China—then an economic backwater with an economy that was one seventy-fifth the size of our own. By letting Communist China into the World Trade Organization, these elites exposed millions of workers to cut-rate Chinese labor and cheap products. At the time, the people who made key decisions about American economic policy believed that the end of history had arrived, and that our primary task as a country was to integrate every society on earth into the global market. We believed that even the worst, most repressive regimes in the world—including Communist China, which had murdered thousands of its own citizens in Tiananmen Square—would magically become liberal democracies if we allowed them to get a little rich off the free market.

More than twenty years later, this decision now shapes the way our most powerful corporations behave. It has drastically affected what products are available for us to buy here in the United States. In the early months of the Covid-19 pandemic, for instance, many Americans were shocked to learn that a vast amount of the world's supply of N-95 masks were manufactured in China, as were the basic ingredients of several key medicines, like penicillin or blood pressure medication, that Americans relied on to get through their daily lives.

Now more than ever, the decision of our elites to allow Communist China into the WTO is acknowledged to be a critical mistake.

The results are in—they've *been* in for years, in fact—and they're not good.

Five years after China became a full member of the WTO, our trade deficit with them had nearly tripled. Before China joined the WTO, the United States was the largest trading partner of 152 countries in the world. Today, we are the largest trading partner of only 57, and China is the largest trading partner of 128.

At the time, our elites were operating on a flawed assumption, namely that global economic integration—as a central tenet of an agenda of liberal openness—was more important than anything else. They believed it was more important than dignified work for Americans. More important than our ability to make things. And more important than our national security.

The assumption was perfectly summarized by President Clinton when he said, "By joining the WTO, China is not simply agreeing to import more of our products, it is agreeing to import one of democracy's most cherished values—economic freedom." The results were predictable.

Instead of importing "economic freedom," as President Clinton predicted, what we exported was our industrial strength. The result has been an economic, social, and geopolitical disaster. Tens of thousands of American factories disappeared, and an estimated 3.7 million American manufacturing jobs went with them.[1] As a result, many of the communities that grew up around those factories became hollowed out. Think of the rust belt, for instance, or the many small towns in the American Midwest that had once supplied the world with steel, textiles, and tools, or the textiles and furniture industries of the upper South. Many of these towns saw sharp declines in population in the early years of the twenty-first century.

Most identify this as the Hillbilly Elegy phenomenon. Before being elected to the Senate, J. D. Vance did the nation an incredible service by breaking down the root cause of despair and devastation in the rust belt.

"Not having a job is stressful," he wrote, "and not having enough money to live on is even more so. As the manufacturing center of the industrial Midwest has hollowed out, the white working class has lost both its economic security and the stable home and family life that comes with it."

But most in the media conveniently ignore the fact that this didn't affect only rural white communities. Black communities stretching from once-prosperous cities like Baltimore to smaller southern towns like Charleston were similarly decimated. It speaks to the widespread destruction these wrongheaded policies inflicted on Americans of all races. Equally stunning is how little you hear Democrat politicians—the ones who have run these cities for generations—talk about the issue. Neglect isn't a strong enough word. It is really a dereliction of duty that deserves a national response.

In 2020 I was preparing to give a speech at Florida A&M University, a historically black university founded in 1887, on this very topic. It was to be the beginning of a multiyear effort to force leaders of both parties to grapple with this failure. Then the pandemic shut down the world. I posted my would-be speech on Medium in May 2020:

> *Between 1960 and 2010, research shows that those urban centers that have experienced the highest losses in manufacturing employment have also "experienced the smallest growth in average wages for black men."*
>
> *The result was a joblessness crisis for African Americans that has now extended across generations and made it harder for America to close the racial wealth gap.*
>
> *We see the effects borne out in drops in black wages and employment rates, as well as a corresponding rise in poverty. Between 1960 and 2010, economic statistics lay bare the material decline. In those fifty years, our society saw an increase of 8 percentage points in poverty rates for black women, a 13.3 percent decline in wages for black men, and a 5.6 percent decline in employment rates for black men. During that same half-century, for men the racial wealth gap increased by 12 percent, and the employment gap increased by 3.4 percentage points.*
>
> *The unbalanced composition of our economy that we have inherited from these decades of misguided economic thinking today bears down with particular force on black Americans. No American should be forced to suffer these indignities.*

For those who stayed behind in these forgotten communities, deaths of despair became more common than ever. In recent years an opioid crisis has emerged, its geographic hotspots corresponding directly to the very regions most impacted by the loss of factories and jobs. This loss has also contributed to historic declines in men's labor force participation, wages, and even marriage. Recently, it was revealed that black Americans now have a higher opioid overdose rate than white Americans.[2]

Today, many Americans—particularly those who live in areas that have been hollowed out by the loss of manufacturing jobs—believe the American Dream is no longer in reach. The dream my immigrant parents and millions like them achieved here in this country. Not a dream of becoming rich or owning a lot of things, but a dream of having a stable and dignified job that allowed you to get married, start a family, own a home in a safe neighborhood, retire with dignity, and leave your children better off than yourself.

Other than some brief periods, like the time my mother worked in the manufacturing plant in Hialeah, my parents didn't work in the manufacturing sector. But they did work in an economy that was balanced. It had a vibrant manufacturing sector buoyed by high productive jobs. It had a financial sector that played the simple but necessary role of connecting people who had money with people who needed it to execute good ideas. And it had a service sector that provided services throughout the economy. It all worked well together and was a sustainable, complementary model.

In recent years, it's become fashionable to pretend that the loss of American manufacturing was a historical inevitability—that increased global communication was always going to lead to factories moving overseas where people were willing to work for lower wages. This is false, as anyone who studies the history behind this belief can see. It's also become fashionable to pretend that manufacturing should no longer play a major role in the American economy.

During the 2016 campaign, when Donald Trump tapped into the frustrations of people who'd lost their jobs and livelihoods over the past twenty years, Columbia Business School published an article titled "Clinton or Trump, Manufacturing Jobs Unlikely to Return."

Fortune magazine published a similar article explaining "Why Trump, Clinton Can't Revive Manufacturing." The prevailing idea seemed to be that the jobs we had lost were gone forever, and that we should cut our losses and move on.

In times of relative prosperity, we tend to believe that the things that use to matter—making things, looking out for our fellow citizens, and shoring up our supply lines—no longer matter. We allow ourselves to believe that an expanding stock market is the same thing as a thriving economy, and that the success of so-called tech companies like Uber, Meta, and Google will eventually trickle down to the working-class Americans who drive their own cars, even though it didn't work out that way during the first tech bubble twenty-five years ago. And in the years since China joined the WTO, most people didn't even blink as finance became a larger share of US corporate profits than manufacturing.

But this signaled an important change in the American economy. It was the moment when financialized paper wealth—that is, wealth that is created on account of a complex system of finance rather than through traditional productive activity—began to overtake the classic American system. Since then, our economy has never been the same.

The American System and Its Discontents

Since the founding of our nation, economic independence and a robust manufacturing sector have allowed this country to thrive. The great statesmen of our past have always understood this. More importantly, they have understood the danger of depending on other nations for our prosperity.

Take Alexander Hamilton, for instance, who served in the colonial army during the American Revolution. During the war, Hamilton saw firsthand how difficult it was to secure tents, uniforms, gunpowder, and blankets for his troops. At the time, the colonial army was dependent on an almost nonexistent manufacturing sector; many of its guns and uniforms came from Europe. During a single winter in 1778, more than 2,500 men—a number that represented one quarter of the army—died from "disease, famine, or the cold," a direct consequence of not having enough factories to produce supplies.

So in the early days of the Republic, Hamilton and President George Washington crafted economic policy that stressed the importance of making products right here at home. As treasury secretary, Hamilton wrote in his foundational 1791 Report on Manufactures of the need "for the United States to consider by what means they can render themselves least dependent on" foreign powers through the "particular encouragement of manufactures in the United States." According to Hamilton, "not only the wealth but the independence and security of a country" were "materially connected with the prosperity of manufactures." America "ought to endeavor to possess within itself all the essentials of national supply," including for defense.

Through the adoption of policies such as Hamilton's, the United States began to lay the groundwork for an economy that would eventually allow "a nation of four million people," in the words of Michael Lind, "mostly farmers and slaves, inhabiting a miscellany of former British colonies along the Atlantic rim of North America" to become "an economic and military colossus with a continental territory inhabited by a population that was predicted to grow to as much as half a billion by 2050."

But it didn't happen without a great deal of careful work from other talented statesmen, business owners, and ordinary citizens. One of the most prominent was Henry Clay, a statesman and senator from Kentucky who continued the work of men like George Washington and Alexander Hamilton, pushing the United States even further along the road to economic independence and prosperity. By the early 1800s, when Clay was serving in the Senate, the British economy was growing rapidly with the help of new machines that made manufacturing faster and cheaper than it had ever been before.

In response, Senator Clay proposed what came to be known as the American System, which called for a federal policy of protectionism and support for young industry in the United States. Under this system, the federal government raised tariffs on foreign goods, spent federal funds to improve infrastructure, and created a sound national financial system that has endured, at least in principle, to the present day.

In short, Henry Clay's vision for American economic policy consisted of two pillars. First, he didn't want American prosperity to depend on any other country. Second, he wanted ours to be a diverse industrial economy. That vision, and the work of those who carried it to the present day, has literally steered the course of history. It was this vision that allowed the United States to become the world's greatest economic power by the late nineteenth century, and what allowed us to maintain that supremacy despite wars, market crashes, and thousands of other domestic issues.

It also helped us to prevail in global conflicts. In the early 1940s, when the United States joined World War II, our armies were not caught without supplies the way they were in the late eighteenth century. By then we had built a system of factories and supply lines that were quickly modified to produce all the clothes, provisions, and munitions that were necessary to support our soldiers abroad. Our industrial capacity and diversity became the war machine that tipped the scales in that conflict, leading to a period of American dominance that would continue for decades.

But the real boom came after the fighting was done. In the decades following World War II, the same factories that had supplied weapons and supplies during the war transitioned back to making domestic products with a renewed fervor, hiring more Americans than ever and ramping up production to meet the needs of soldiers who were returning home. In doing so, they transformed every corner of this great nation—even my hometown, which is most often associated with beaches and sunshine. By the time my family arrived in the 1970s, the textile and apparel industries in Hialeah had become a point of pride not only for the people who worked there but for everyone in town who was able to share in the prosperity created by these factories.

But it wasn't just Hialeah. In fact, compared to most major manufacturing towns, my little corner of Miami was just a blip on the radar screen.

Throughout the twentieth century, America built thousands of manufacturing plants, most of which were in small, formerly depressed areas in the Midwest. According to an article in *Industry Week Magazine*, these included "food processing plants, auto manufacturers, textile fabric mills, cut and sew apparel mills, paper mills, foundries, hand tool manufacturers, major appliance manufacturers, machine shops, and many others. . . . When these plants were built, whole communities formed around them providing good paying jobs for millions of people without college degrees, as well as jobs for all of their supplier companies and the merchants in the communities."

Suddenly, more people than ever had a chance to achieve the American Dream. They didn't need to go to college to do it, and they didn't need to go to advanced schools for a decade to gain the necessary skills. They didn't spend their lives trying to pay off $100,000 in student loan debts they accrued preparing to enter the workforce. In most cases, they could learn on the job. When they walked through town, the fruits of their labor were everywhere, even if they didn't notice it at the time. A steelworker from Monessen, Pennsylvania, could go to any construction site on the East Coast and know that the beams being used probably came right from his hometown. Garment workers from manufacturing towns all over the country could walk into almost any department store and see the *Made in America* labels on their clothes.

But as the euphoria of the 1990s set in, different kinds of politicians and business leaders began making decisions about the future of the United States. They began to forget about the millions of people who relied on the steady incomes that our robust manufacturing sector provided. Slowly, in the boardrooms of think tanks and the classrooms of Ivy League universities, a consensus emerged: history, in the words of Francis Fukuyama, was over, and the only thing left to do was spread liberalism—and with it the hallmarks of globalized American economics and progressive values—to every corner of the world.

It was a simple idea. America was the richest country in the world. From a foreign policy perspective, the idea was that America could easily afford to give away some of our industry to foreign nations through generous trade deals in exchange for making them closer allies with us. On its own terms, at least this was a clear tradeoff. But neoliberalism in economic and foreign policy, which can be roughly defined as a belief in free markets and deregulation above all else, skewed the choice in one direction. Neoliberal economics said that America would actually *gain* from giving away our industry to foreign nations, because consumer products would be cheaper and American workers could give up their manufacturing jobs for supposedly more "productive" jobs in finance and digital technology. Neoliberal foreign policy said not to worry about it; after all, the US dollar was still the world's reserve currency, and foreign nations would ally with us simply because they would become wealthier. Combined, it was a recipe for disaster. After China was admitted to the WTO, it was a disaster on the grandest scale.

At first the changes were relatively minor. A few companies moved key operations overseas, allowing them to take advantage of the cheap labor available in countries such as India and China. This practice, known as outsourcing, had become increasingly common in the late 1980s, and accelerated in the 1990s. Soon, with the passage of legislation like the North American Free Trade Agreement (NAFTA), companies realized they could also take advantage of a practice known as "offshoring," whereby they closed down entire factories in the United States and moved them to places like Mexico.

All of this set the stage for allowing Communist China to join the WTO.

Looking back, it might seem as if these developments occurred with little resistance from the people who were going to be hurt the most. But they didn't. Time and time again, our elites were warned of the potential consequences of their actions.

In the early 2000s, for instance, many publications pointed out the dangers of allowing Communist China into the global economy. First, they pointed out that our trade deficit with China was already unbalanced. According to an article published by the Economic Policy Institute at the time, the US had "imported approximately $81 billion in goods from China" in the year 1999 and "exported $13 billion—a six-to-one ratio of imports to exports that represents the most unbalanced trade relationship in the history of US trade. While exports generated about 170,000 jobs in the United States in 1999, imports eliminated almost 1.1 million domestic job opportunities, for a net loss of 880,000 high-wage manufacturing jobs."

The trend was clear. With every month that passed, the trade deficit between the United States and China was growing. More American jobs were being lost to cheap Chinese labor. And this was *before* China entered the WTO. Once they did, according to many experts, the agreement would "lock" that bad relationship into place, "setting the stage for rapidly rising trade deficits in the future that would severely depress employment in manufacturing, the sector most directly affected by trade."

The outcry did not come only from policy journals and the editorial pages of newspapers. In the late 1990s, many Americans from all classes and backgrounds were concerned about the threats posed by rapid globalization. They knew that allowing for unregulated competition from foreign workers would have devastating effects on workers in the United States.

And the workers knew it, too. But anyone who expressed doubt about the rampant pace of globalization was portrayed as naive, radical, or unwilling to accept the inevitable. The only opposition in Congress was from a small minority of elected officials from the left end of the Democrat Party and the right end of the Republican

Party. But the criticism was vocal—and with the benefit of hindsight, it was true. One of the most vocal opponents of giving normal trade relations to China, North Carolina Republican senator Jesse Helms, said, "We Americans stand for something—something other than for profits. . . . The Chinese government continues to repress, to jail, to murder, to torture its own citizens." On the left, only a coterie of unions and environmentalists opposed admitting China to the WTO, along with a little-known socialist congressman from Vermont named Bernie Sanders. "Both political parties are way out of touch with the American worker," Sanders said at the time. He went to blast the idea of "allowing large, multinational corporations to throw American workers out on the streets, move their plants to China, and bring their products back into this country tariff free."

It would take almost thirty years for the objectors to be proven right. In the meantime, American elites, driven by a bipartisan consensus that global economic integration was inherently good, ruled the day.

Senator Joe Biden, who had been in the Senate for nearly three decades by that point, was a supporter of integrating with China. For years, Biden had adhered to a flawed post–Cold War view of the world. He, like many of his colleagues, believed that if we could only spread the ideology of free markets around the world, then even the worst regimes on the planet would adopt the principles of liberal democracy and become more like the United States. He especially believed that this would happen in China, a country where his family is currently engaged in several business deals.

In his first speech in the Senate, Henry Clay warned that some "have been engaged, to overthrow the American System, and to substitute the foreign." He suggested that while they were "long a resident of this country," they have "no feelings, no attachments, no sympathies, no principles, in common with our people."

Those words were spoken in 1832. But they have taken on a new relevance today.

Flawed Assumptions

Shortly after Joe Biden was elected president, he released a list of his cabinet picks. After taking a quick look, I sent a tweet that was apparently controversial. "Biden's cabinet picks went to Ivy League schools," it read, "have strong resumes, attend all the right conferences. And they will be the polite & orderly caretakers of America's decline."

The tweet definitely hit a nerve. It did so because it was so true. From the list of names, it was clear that Joe Biden was nominating the same people who'd gotten us into so much trouble in the first place—the ones who worshipped a post–Cold War ideology that had been proven false time and time again. Still, these people cling to the same assumptions that led to our blunder with China and the WTO. And if we're going to have any hope in the future of correcting this blunder, it's important to understand what these assumptions are. This way, we'll be able to adopt better ones in the future.

The first flawed assumption is the belief that Americans are primarily consumers—that our primary economic identity is not that we are workers, parents, heads of households, or members of a community but that we are shoppers. If you believe that we derive our identity and our happiness from the things we buy, and not from work, children, the family, then it becomes easy to justify opening up to China in exchange for cheaper prices. Life, liberty, and the pursuit of cheap consumer products.

But we all know—as did our nation's founders—that so much of what gives life meaning and purpose isn't how many things we can buy or how many things we own; it's the time we spend and the things we do with our family and our communities. That requires the stability that comes from good jobs.

To view Americans solely as consumers ignores the dignity that comes with work. And it ignores how corrosive it is to the individual and ultimately to a community when good jobs are no longer available.

It leaves you with an unemployment rate, for example, that drops consistently below 5 percent, not because more people are working but because more people have given up looking for dignified work and dropped out of the labor force.

It leaves you at the mercy of any disruption in the supply chains. Because when shortages make it harder to find what we want to buy and inflation makes something that was once cheap more expensive, you still don't have the jobs, you still don't have the dignity that comes with work, you still don't have the industrial capacity that comes with American labor, and now you no longer have the cheaper prices. That's the predicament we unfortunately have found ourselves in repeatedly over the last several years.

And eventually, if cheaper prices at the store are the result of sending the job you once had to a cheaper worker in another country, you're inevitably going to face widespread anger and despair.

And still, many don't get it, ignore it, or hope we overlook it.

When we have the Chamber of Commerce and the Business Roundtable petitioning the Biden administration to lift Trump's tariffs on China—or when ninety-one senators vote to cut tariffs on Chinese goods, as they did as recently as 2020—it shows you just how deeply embedded flawed thinking still is in our country. I was proud to join three of my Senate colleagues—Josh Hawley, Tom Cotton, and, yes, Bernie Sanders—in voting against cutting tariffs on Chinese goods. It's alarming that there were only four of us.

The second assumption that underpinned the consensus that led to this decision was that the stock market and corporate profits are the same thing as real economic growth and innovation.

For twenty years now, presidents from both parties have pointed to the stock market as a scoreboard indicating whether we are winning or losing economically. A thriving stock market is not a bad thing, nor is it unimportant. But our economy is a lot more than just the stock market. And most of the time, whether the market closes up or down, has zero correlation with how the economy is working for our country, for our families, and for our communities. Today it has become a form of high-stakes gambling.

Over the past two decades, the stock market has gone up 120 percent when adjusted for inflation, but "middle-income" Americans have only seen marginal growth of about 6 percent over that same period. And lower-income Americans have actually seen no growth at all.

And when did this trend take off? When did all of this begin to happen? It coincides almost perfectly with the collapse of American manufacturing after China entered the World Trade Organization. Wall Street and corporate America discovered that they could now make the same products, or something new they invented, at a lower cost by using cheaper workers in Chinese factories.

The lower cost of manufacturing meant not just cheaper prices but also larger corporate profits because of lower costs. And larger profits meant greater returns for shareholders, but it didn't necessarily mean greater prosperity for working Americans. As Oren Cass has pointed out in a piece published by American Compass, even Adam Smith's theory of capitalism doesn't presuppose some invisible hand magically making sure that larger profits are tantamount to societal betterment. In fact, Smith's great work *The Wealth of Nations* makes it clear that capitalism works only when profit maximization is aligned with the national interest. As Smith writes, "Upon equal, or only nearly equal profits, therefore, every individual naturally inclines to employ his capital in the manner in which it is likely to afford the greatest support to domestic industry, and to give revenue and employment to the greatest number of people of his own country."

There is nothing inherently wrong with greater profits for corporations and better returns for shareholders. But those things are meaningless if our country and our people are not stronger and more resilient as a result. Lower prices alone can never make up for the loss of the stability and dignity that comes from a good paying job. Greater returns on the stock portfolio of investors can't make up for the closure of a factory that left behind a hollowed-out community. And record corporate profits can't make up for the insecurity of a nation that can't make masks during a pandemic or produce the active ingredients in our most basic medicines.

When you have an economy where wealth is being generated in a way that is divorced completely from the well-being of a people or the security of a country, it foments discontent and the resentment that Marxists always seek to exploit. It gives an opening to argue that capitalism is inherently unfair and repressive. And it creates an opportunity to argue that the time has come to abandon free enterprise and empower the government to "Build Back Better."

And when wealth is generated by sending our jobs and our manufacturing capacity to a country that seeks to rise at our expense, it leaves a relationship with a zero-sum game, in which either they win or we do. It leaves you a nation vulnerable to them, whatever disruptions they seek to create, and to any coercion they seek to pursue.

The third belief behind the flawed bipartisan consensus was that if our companies sent our factories and jobs to China, it would give America more influence over China and the world. The failure is so obvious that Richard Hass, one of the chief cheerleaders of American globalism, acknowledged last year that "integration, which animated decades of Western policy toward China, has also failed." As a hedge, he suggested it was debatable "whether the flaw lies with the concept of integration or with the manner in which it was executed."[3] I am sure the execution was flawed, but perfect execution of a naive concept will still produce negative results.

I know it sounds silly today, but not long ago there was a catchy talking point that was popular in college economic courses and conferences and soundbites. It came from Thomas Friedman, the *New York Times* columnist who shows up at every fancy gathering of our nation's elites: "No two countries that both have a McDonald's have ever fought a war against each other." The first McDonald's opened in the Soviet Union in 1990. A Ukrainian boy commented at the time, "I thought [America] wanted to launch nuclear rockets at us, but they gave us McDonald's and peace instead." America has always sought peace, because we are the only great nation that has never sought to be an empire. But the presence of McDonald's in both Russia and Ukraine did nothing to dissuade Vladimir Putin from launching his invasion. It turns out that Vladimir Putin and his friends in the Kremlin have very different values than Thomas Friedman and his friends in Aspen.

Twenty years ago, the world's most powerful companies were American. And so the belief was that this new world centered on global commerce would inevitably lead, not just to more McDonald's, but to greater American influence as well.

But that's not how it turned out. We soon learned that you could eat Big Macs and still view the country that invented them as a rival to be defeated. And we learned that when American companies are forced to choose between what's good for America and bigger profits, they will usually pick bigger profits.

The Chinese Communist Party never wasted any time believing that the "end of history" had arrived. They saw, and still see, history as thousands of years of greatness interrupted by a century of shame and humiliation at the hands of the West. And they viewed it as their destiny to become the world's preeminent power at America's expense. For a time they chose to "hide their capacity" and "bide their time." But by 2008 they felt strong enough to no longer pretend.

Over the past twenty years the Chinese Communist Party has become more repressive than ever. The party censors the speech of journalists and prevents its citizens from seeing any information that is even remotely critical of the government. China has also begun cracking down on its population of Uyghur Muslims, an ethnic group historically dominant in the western province of Xinjiang. As of this year, more than a million men, women, and children from the Uyghur population have been forced into internment camps by the Chinese government. Within these camps, they have been raped, tortured, and stripped of their identities and their faith. Since 2013, when President Xi came to power, other ethnic and religious minorities have been placed into similar "re-education" camps.

Tragically, American corporations used these camps to source materials to make their products, effectively profiting from the gross human rights abuses of the Chinese Communist Party. That wasn't only immoral; it was deeply dangerous to America and its working families.

In 2018 I coauthored the Uyghur Forced Labor Prevention Act with Senator Jeff Merkley of Oregon. The goal was to make sure products produced with the forced labor of the Uyghurs would not be permitted to enter the United States. After a long fight with corporate America and the Biden administration, that bill finally became law on December 23, 2021. It was finally a change to federal law that signaled some things are more important than making money in China. But we cannot afford to pat ourselves on the back for a job well done. That same principle needs to go well beyond T-shirts made with slave labor.

For years, I've tried to warn about the dangers posed by the Chinese Communist Party. In 2019 I released a report titled "Made in China 2025 and the Future of American Industry," which laid out my primary concerns and put forward a plan to address them.

Unfortunately, many in our government have been too slow to respond to this threat. Many do not seem to care at all. Wealthy progressives in government and industry have even welcomed the changes, celebrating an increasingly interconnected and multicultural world. After all, elites were still doing well; they married, and continue to marry, at historic rates, and raise children in stable homes and neighborhoods. Some in flourishing Silicon Valley would even look at hollowed-out communities and individuals facing job loss and flippantly tell them to "learn to code."

For our own part, conservatives have too long only pointed to cultural decline as the reason why these places are not flourishing. If the best way to have a stable society is to raise children in a stable, two-parent household, then a community that lacks these stable households is likely to struggle. And these conservatives would say that cultural values that prioritize marriage and family should hold regardless of one's financial situation.

There is a lot of truth to that. According to the data, marriage and childbearing have declined across incomes and backgrounds. Hallmarks of community life like church attendance and participation in charitable activities have also declined in regions of all shapes and sizes. America has undergone a significant cultural change, and the causes of it are complex. There is only so much that national politics can do, or that it would be appropriate to do, to reverse these changes. My wife and I often pray for a revival of religious faith in our country, but that is for God to decide, not politicians.

However, at the same time, in visiting these communities, it's just common sense that their economic decline has affected community life. Young men who would have previously worked in productive industries are increasingly out of the workforce and idle. To be sure, the right thing for these men to do would be to get a job—any job—and work to support themselves and their families. But we shouldn't pretend that the decline of well-trod paths to work in well-known industries for young men has not had a cultural effect.

Cultural decline isn't an issue that is isolated from economic decline. Dignity comes from a job, and so a difficult economy is a drag on the dignity of the workers in that economy. For a long time it has been impossible to acknowledge this inside America's conservative movement. I'm glad those days are changing, and I hope I've played a role in breaking that taboo. A healthy economy isn't sufficient for a healthy society. But we need to stop pretending we can have a spiritually healthy nation if our economy is leaving millions of people behind.

Chapter 2

PAPER WEALTH

The Divide

When you run for president of the United States, the demands on your time are incalculable. Many of those days are just a blur at this point, but there are some memories I will carry with me for the rest of my life. Some are scenes that might seem familiar to any American family—the snowman that Jeanette, myself, and all four of our kids built one morning in Hudson, New Hampshire, for instance. I also remember a young boy who followed me around at one campaign event with his digital camera, proudly recording the whole thing for posterity. I'm sure I'll never live down the moment when, playing catch with the nephew of my Iowa chairman, I threw the ball right through his hands and knocked him to the ground.

Most Americans assume that a big part of running for president is meeting with big donors and trying to win their favor. Many political consultants say that "if you aren't running for president five years before the election, you aren't going to be prepared."

There was a time when this perspective was true. From Reagan to Romney, the GOP had a tendency to nominate the next guy in line. All the next guy in line had to do was convince a relatively narrow group of people that he could do the job. Political scientists call this the "party decides" model—a system where, intentionally or not, the election is decided by the top consultants, donors, and elected officials in the party. There certainly have been elections that worked this way. But they occurred years ago, back when politics played a smaller role in our nation's emotional life. In those days, people tended to find community at their local church, tribalism at their favorite football team's home games, and entertainment from the Hollywood studios. The stakes of our elections did not seem existential the way they do today.

Oddly enough, this began to change about seven years ago—right when I decided to run for president. By that time, the reelection of President Obama had triggered an all-out assault against our traditional American values. The backlash was beginning to grow against both Democrats and Republicans. People wanted somebody who would fight a system they viewed as corrupt and stand up for them. They had an impression that the powerful and well-connected had too much access in politics, and that this trend had caused elected leaders to lose their way.

After spending an inordinate amount of time running around the country to raise money for my campaign, I can say people had every reason to have this perception and be furious about it. The finance part of running for president entailed attending many parties and dinners with the financial elite of this country. For the most part, these events were fine. The people were perfectly nice and some became genuinely good friends of mine whom I value to this day.

But when you attend these events, especially if you come from a working-class community, you can't help but be reminded of a sharp divide that exists in the United States—not between left and right, or even the rich and poor. This divide ran much deeper than any of those, and it has been growing exponentially for the past thirty years.

I'm talking about the divide between a small, isolated group of largely coastal Americans—what some have labeled our nation's ruling class—and everyone else. The ruling class is the collection of hyperwealthy financiers, journalists with the largest megaphones, celebrities who can shine the brightest spotlights, and public intellectuals paid enormous sums to speak at Aspen or Davos or the other conferences at which the previously mentioned groups congregate.

They aren't by any means evil people, but they are extraordinarily out of touch with the day-to-day experiences of their fellow Americans (though many members of our nation's ruling class would object to the term "fellow Americans," as they think of themselves more as "citizens of the world"). If you think I'm joking, by the way, take a look at Stanford University's Elimination of Harmful Language Initiative, which includes *American* in its list of harmful words.[4]

To be clear, the divide between right and left is real, and it is growing deeper. But as that fight plays out in school board meetings and committee hearings across the country, members of our ruling class circle above like vultures. For the ruling class, the cable news battles that play out between left and right are an easy distraction from the bank heist they are carrying out on our country.

During my other campaign travels, it was clear that issues like immigration, the opioid crisis, and trade with China did, in fact, matter to voters. Over the course of several months, I sat with union representatives on either coast, shared meals with local factory workers in Dubuque, Iowa, and met thousands of young people eager to get involved in politics on college campuses all over the country. I heard from working-class families who were struggling financially, and I listened to stories about people whose lives still hadn't returned to normal after the financial crisis of 2008.

I saw firsthand that despite a stock market that was slowly creeping upward, working people in this country had not experienced the recovery that many at the top were enjoying. This was especially true in factory towns that had been devastated by offshoring and the slow, steady hemorrhaging of jobs that came in the aftermath. It was also true of two-income families who could no longer afford the cost of childcare. To these people, it did not matter that major American companies were paying record dividends to shareholders; they still had record-high levels of credit card debt and more trouble than ever trying to buy houses. Many of them reported feeling hopeless.

Most importantly, I saw that a majority of working-class people in this country—the people who were doing the same types of jobs my parents had done decades earlier, often in the same places—did not feel like members of either political party had much to offer them. From where they were sitting, it looked an awful lot like Republicans only cared about cutting taxes for the rich, and that Democrats were only interested in raising taxes, giving out government assistance, and blathering on about identity politics. These things might have seemed important in Washington, but they were laughable in the cities and towns I visited during the campaign. Among other things, I found that unemployed steelworkers in rural Pennsylvania didn't take kindly to being told by affluent East Coast liberals that they had "white privilege," or that the exhaust from the cars they drove represented a greater threat to society than, say, rampant globalization and all the disastrous domestic consequences it brings.

The stories these people told me influenced my legislative priorities in Washington; they also shaped the speeches I would give and the issues that would become central to my campaign. Looking back, the months that I spent traveling around the country were among the most fulfilling of my life, and they continue to shape my work in the Senate today. When I find myself stuck in the middle of writing a bill or debating how to vote on a piece of legislation that someone else has proposed, I'll often think back on the months that I spent campaigning for president. If I can see myself standing in front of a small crowd in Iowa, New Hampshire, or West Virginia and telling them with pride that I supported the legislation in question, I'll vote for it without hesitation. If I can't see myself doing that, I vote no.

But as I mentioned, running for president, I was told, was about much more than just laying out the most compelling vision. I needed to raise a lot of money, and as a result I spent a great deal of time during the campaign flying to private dinners and cocktail parties, where the elite megadonors of the Republican Party stood in judgment of candidates vying for the nation's highest office. As you can tell, I'm increasingly skeptical of the extent to which these people matter when it comes to winning a campaign. The reality is, the gulf between the Republican Party's donors and voters has never been greater, and ultimately (and thankfully), the voters hold the power. But if nothing else, I came away from them with a better understanding of why our donor class believes what they believe and why their advice to lawmakers—often treated as gospel by some in my party—is so often wrong for the country.

When the town-hall rallies in New Hampshire or Iowa were done for the afternoon, I'd get on flights to New York City or some other urban enclave and prepare to rub shoulders with the financial elite of this country. My staff often had to drag me off the rope line.

On our way to the airport, they would remind me of who cared about what in the room I was about to walk into. There is a hedge fund manager who has supported Democrats in the past, for instance, and he really wishes I wouldn't talk about abortion as much. There's a bank executive who's really worried about Trump's rhetoric on immigration and wishes I hadn't pulled the rug out from under the Gang of Eight bill a couple years earlier. One thing I rarely heard was that this Fortune 500 company CEO is really interested in moving supply chains back to America. Or that Wall Street guy is very concerned about China's exploitation of our capital markets.

That makes sense, of course, because for these people, the economy is working just fine. For this group, those big economic numbers are reality, a measure of wealth and success. When the stock market goes up, they are happy. Even in the years after the 2008 financial crisis, when the major banks who'd caused the crisis were bailed out, it was this group that benefited most. Many of them didn't seem to care that the government didn't bail out homeowners who'd suffered during the crisis, or that historic numbers of Americans had lost their jobs in the aftermath. In some cases, they had deep ties to the banks whose risky financial maneuvers had led the housing market to collapse in the first place.

For the past thirty years, the financial industry that nearly brought this country to its knees in 2008 has come to dominate a larger and larger piece of the American economy. Business profits have become increasingly estranged from production and employment. This is mainly driven by large transnational corporations. Many of these corporations are now using our country's resources to speculate on financial assets, including their own share prices. Rather than engaging in real production and innovation with workers here at home—the production that delivers widely shared prosperity—they have sought to reduce their domestic labor costs. This strategy is damaging not only to the American worker but also to the competitiveness of American industry.

We are cutting off the branch on which we sit.

In early 2019 I released a report titled *American Investment in the 21st Century*. The report argues that underinvestment in America's economy is driven by the consensus that the goal of business enterprise is to maximize financial return for shareholders. It is easy to see how this belief would lead to lower physical investment. Returns from financial engineering are easier, quicker, and more certain for shareholders than long-term investment in the capital- and labor-intensive creation of actual goods and services.

So that is what we have pursued.

As a result, American oligarchs from both political parties have gotten historically rich and further removed from American life, which is collapsing as our economy becomes distorted and highly focused on financialized paper wealth. The decision to bail out financial institutions in October 2008 revealed the disastrous disconnect between finance and the real economy, and the consequences of making finance America's leading industry. The financial crisis revealed something more than a lapse in regulatory oversight. It revealed the extent to which working Americans' jobs, savings, and ultimately communities are subject to an abstract numbers game on Wall Street and across corporate America.

This doesn't mean that the financial services industry is bad, or that we all need to start following the lead of socialists like Bernie Sanders when it comes to the market. So that there's no confusion, let me state very clearly what I believe: The finance industry is a necessary and noble profession that plays an important role in the economy, connecting people who have money with people who have great ideas that need funding. The people who work in finance are not evil, and the people who start businesses are not any better. But finance as an industry has grown beyond the service job I've described above into an end in itself. That is why we need to take a long, hard look at the shape of our economy and figure out how we got here. And we need to do it before the risky, overly complex nature of our financial system leads to another global meltdown.

Which, if recent trends are any indication, might happen much sooner than you might think.

Financialization

The beating heart of American capitalism has always been businesses making stuff. The heroes of American capitalism are the doers, the people who innovate and create something—from the small business owners to the corporate leaders who feel an obligation to invest in America.

For most of American history, the role of American finance was simple. Banks lent money to help grow the economy. When a man who wanted to open a diner or a flower shop walked into a bank, he could be reasonably certain that as long as he had good credit and a strong business plan, he'd be able to secure a loan and open his business. Banks were responsible for "allocating capital." They would take the money that had been deposited into the savings accounts of normal Americans and loan it out to people who wanted to start businesses, thereby greasing the wheels of our economy and helping to create jobs and support businesses on Main Street.

It wasn't sexy or flashy. It was the slow, steady pace of sustained growth—investments that built not just a return on paper but the foundations of strong families and communities across the country.

What the banks would *not* do—at least not on a large scale—was take the debt that resulted from these transactions, bundle it up into complex financial instruments, and sell those financial instruments to other banks in the hope that they'd make massive profits on interest payments and transaction fees. And businesses would not make finance like this their primary source of wealth; they'd rely instead on investing in their line of business.

I'm sure most people have seen the film *It's a Wonderful Life*, which was written when many Americans were still dealing with the fallout from the Great Depression. Most people probably remember the famous scene when George Bailey reminds the angry, panicked citizens of Bedford Falls that although their money is not "in" the vault of the bank, and therefore cannot be withdrawn all at once, it *is* invested in the community. There's some drama in that film, of course, but the picture it paints of a staid, quiet banking industry is relatively accurate.

During this period, income inequality decreased. Banks did what they were designed to do: they took the savings of Americans and invested those savings in ways that helped the economy of this country to grow. The decades following the Great Depression saw a tremendous amount of prosperity for working men and women in this country, my own family included.

Of course, I'm not saying it was some happy black-and-white world where everyone always got what they wanted. People failed all the time. Before I was born, for instance, my father attempted to use his experience as a bartender to open up his own small businesses in Miami; he tried a vegetable stand, a dry cleaner, and a small supermarket, but all of them closed within a few months of opening. Years later, my mother would tell me that he was always too generous with customers to run a business—that he would give away items to people in need and allow them to go months without paying for them.

Eventually he worked several good-paying jobs for businesses owned by other people, which is how most people in the United States make a living. But the owners of the businesses he and my mother worked for benefited greatly from the overall growth of the

American economy, which was stronger than ever during the latter half of the twentieth century. In my hometown alone, I saw enough new bars, restaurants, and clothing stores to know that people were succeeding all the time. In fact, over the course of the 1970s and '80s, Cuban immigrants like my parents sparked an entrepreneurial boom in Miami, opening so many businesses in town that the effort became known as "the Cuban Miracle."

Living in this area, my father was able to settle in this country with little education and sustain a stable family life with his job. He and my mother owned a home, raised four children, and cared for my grandparents on the annual incomes of a bartender and a maid. We could even afford for my mother to spend most of her time at home when I was young.

To many people, it seemed that this arrangement might always be common—as if the relative prosperity of the working class would go on forever.

But then something changed. In the 1970s, inflation became a huge problem for many Americans. Wages stopped growing for the working class. To fix this, our government began ceding more and more power to the financial sector, which could inject a little life into the economy through inventive financial maneuvers. Throughout the 1980s, which came to be defined by films such as *Wall Street* and its famous phrase "Greed is good," the financial industry began reasserting itself not only as the means by which American business could thrive but as a major business in and of itself.

It's not a bad thing if the financial sector grows, of course. What *was* bad was the financial sector growing even as factory jobs disappeared, inflation soared, and wages stagnated. It signaled a growing disconnect between financial services and the sectors they formerly existed to serve. Finance was becoming the end itself, not just the means.

In 1950, when postwar America was booming, the financial sector made up about 2.8 percent of the economy. By 1980 that number rose to 4.9 percent, and it reached 8.3 percent before the banks nearly destroyed the economy in 2008. A similar trend is clear in corporate profits. The financial sector's share of corporate profits grew from about 10 percent three decades ago to a peak of about 40 percent in the prerecession 2000s, and after the 2008 recession it rebounded to about 30 percent by the mid-2010s.

No surprise, traditional American manufacturers started to chase those returns. In the 2000s, Ford Motor Company made more from the loans it issued than from car sales.

Perhaps the most damning evidence of all is that in the late 1960s and early 1970s, about 6 percent of Harvard Business School graduates went into finance. By 2008 it was 28 percent, and 29 percent in the class of 2018. Only 5 percent went into manufacturing. Some of our nation's brightest, most capable minds were funneled into finance, either because they were seeking a quick return on their massive tuition costs or because it was simply the normal thing to do if you were smart. Regardless, we're seeing that play out today as a nation desperate for engineering talent, manufacturing ingenuity, and technological innovation. Those were areas America dominated in decades past, but no longer do.

Meanwhile, jobs in our communities were disappearing. This was especially true of our manufacturing sector, which continued to shrink as companies moved their manufacturing capacity overseas. For those who still had jobs, wages remained stagnant, while the prices of key household goods went up at a much higher rate every year. Looking back, we can see that these three things—loss of jobs, inflated asset prices, and the stagnation of wages—were a grim warning sign of things to come.

But everyone ignored the warning signs because things looked good on paper. The stock market was going up, and politicians began pointing to that number as an indication of our general economic health. In some ways, they weren't wrong to do so. Beginning in the late 1970s, when individual 401(k) retirement accounts—which are tied directly to the stock market—began to replace company-sponsored pensions, American citizens became more tied to the stock market and less tied to the businesses they worked for.

When the stock market did well, it was generally assumed that they would do well, too.

But there was more to the story. Throughout the 1990s, as jobs were being shipped overseas and our manufacturing sector began shrinking, the financial sector had to come up with more and more creative ways to make the stock market keep rising. Usually, this involved getting bigger and more complex, neither of which they could do without the help of the government.

Luckily for them, the Clinton administration—which was driving the forces of globalization and unknowingly shrinking the size of the real economy every year—was always happy to help Wall Street get bigger. During the Clinton years, more "experts" than ever stepped in to help financial markets grow, almost always in a way that created massive risks for the people who relied on those markets for their retirement savings. One of these experts was Robert Rubin, who served as Bill Clinton's secretary of the treasury. During his time in that position, Rubin worked with his deputy, a young former Harvard professor named Lawrence Summers, to craft policy that seemed to benefit the market (and the people who invested heavily in it) at all costs.

As the economist Rana Foroohar has pointed out in her book *Makers and Takers,* one of their key moves was to adjust tax rates in such a way that people who actually work for a living—which include doctors, lawyers, and lawmakers in addition to plumbers, contractors, and restaurant workers—are taxed at a much higher rate than those who make money by moving other people's money around. We slowly started to chip away at this Clinton-era disparity with our 2017 tax bill, but were limited in what we could achieve because of immense donor pressure. Donor pressure, by the way, caused Democrats to avoid the issue altogether in their 2022 tax hike bill.

Rubin and Summers also pushed for policies that changed the

way corporate CEOs were paid. Under these policies, according to Faroohar, tax-deductible CEO pay would be capped at $1 million, but any "performance-based pay" beyond that—which was typically awarded in stock options rather than cash—would be exempt. It sounded good at the time. "The tax code should no longer subsidize excessive pay of chief executives and other high executives," Clinton said at the time. But his plan had a huge loophole: the personal compensation of CEOs simply became tied more than ever to the stock prices of the companies they led. Rather than incentivizing the leaders of companies to hire more American workers or make products that would add to our general prosperity as a nation, the Clinton administration made sure that they would only have incentives to bring their stock prices higher and higher.

As a result, the frequency of stock buybacks—a move by which a company purchases shares of its own stock on the open market to drive the price up, resulting in greater returns for shareholders—rose dramatically. In 1992 there was less than $50 billion in buybacks; by 1998, that number was nearly $200 billion.[5]

During this same period, the number of complex financial instruments sold among Wall Street banks also grew larger than ever before. Toward the end of the 1990s, in the name of "modernizing" the financial industry, several banks merged to form Citigroup, then the largest financial entity that had ever existed on Planet Earth. Soon other financial behemoths would follow suit.

Perhaps it is no surprise that when Robert Rubin stepped down as treasury secretary, leaving Larry Summers to take his place, he soon became the CEO of Citigroup, taking advantage of the cozy compensation packages he'd negotiated working in President Clinton's cabinet. Summers would come back to the government as President Obama's top economic advisor in 2009. The year before, he made $5.2 million at a hedge fund working what averaged out to about one day per week.

Summers is not alone in driving America over the financial cliff. He was simply doing what everyone else was doing, what everyone assumed was the right thing to do. But he and pretty much everyone else on Wall Street and in Washington during the 1990s forgot that financial markets are only worth something insofar as they create value in the underlying economy. They are supposed to serve the country, not the other way around.

Soon, this approach to public policy and financial services would have disastrous results for the American people.

The 2008 Financial Crisis

By the beginning of the twenty-first century, the financial industry had grown to encompass a much larger portion of the American economy than it ever had before. When my parents were working to support our family, it was generally understood that Wall Street existed to help American businesses thrive. But as time went on, Wall Street became a major business in and of itself—one that created its own products, sold them to a tiny number of other bankers, and made historic amounts of money doing so.

The phrase "Greed is good" became part of the public lexicon thanks to the 1987 film *Wall Street*. But what was seen as distasteful and vile would soon become commonplace as these companies fundamentally changed their mission. Rather than funding businesses on Main Street—which is what it was designed by our founders to do—the financial sector began "securitizing" existing assets such as stocks, bonds, and homes, bundling these assets into complex financial instruments that could be traded as many times as possible. By the early 2000s, many of these instruments had become so complicated that even many of the CEOs and board members of the banks themselves weren't quite sure how they worked.

One of the most famous examples was the mortgage-backed security: a group of thousands of mortgages, many of which were "subprime" (read: extremely risky), that had been bought, bundled up, and offered up for sale by investment banks. At the time, the banks believed that by mixing bad mortgages with good ones, the risk was "diluted," thereby making the security safe to own.

But as the world would soon learn, there was nothing safe about them.

For years Wall Street bankers, in the words of the financial journalist Andrew Ross Sorkin, " 'ate their own cooking'—in fact, they gorged on it, buying mountains of mortgage-backed assets from one another. . . . As a result of the banks owning various slices of these newfangled financial instruments, every firm was now dependent on the others—and many didn't even know it. If one fell, it could become a series of falling dominoes."

Which is exactly what happened.

Beginning in the fall of 2008, as people stopped paying their mortgages, these securities began to fail, and the banks that owned them descended into panic.

For years, almost everyone in the United States had doubted that this day would come. Aside from a few professors and rogue traders who could see that a bubble was about to burst, no one was willing to look the problem in the eye. Speaking in 2007, as the first cracks in the system were beginning to emerge, Federal Reserve chairman Ben Bernanke said that even if the subprime mortgage market did collapse, "the impact on the broader economy and the financial markets" would be "contained."[6]

No one wanted the party to end. For years, the people who worked at these investment banks had been getting historically rich by moving money from one place to another. They packaged the debt of regular Americans—the people they were supposed to serve—and traded it largely among themselves, creating record amounts of "paper wealth." This wealth was not based on anything real. It didn't come from the construction of roads or bridges, and it wasn't tied to the sale of any goods. It certainly didn't create jobs for anyone other than the tiny number of traders and executives who made a collective $53 billion in the year 2007 alone by pushing it around Wall Street.

Underneath all this paper wealth, the real economy was still struggling. Wages for most Americans had not risen since the late 1960s, and to cover for this historic failure, the government had been relying on its old free-market tricks to distract the public. They had lowered interest rates, incentivized share buybacks, and attempted to increase the amount of lending that banks could do. Most crucially, the government had encouraged an "ownership society" under the Bush administration, which was supposed to increase the rate of home ownership in the United States. To do so, mortgage companies drastically decreased the standards for borrowers, resulting in hundreds of thousands of mortgages that were considered "subprime" being sold every year.

Looking back on the period between 2000 and 2007, as the economists Atif Mian and Amir Sufi have pointed out, it is clear that American citizens took on more debt than they had since the decade leading up to the Great Depression. In many ways, the story was the same. Growth in the real economy was slow, but Americans kept hearing that things were good because the stock market was rising. They took mortgages they probably couldn't afford, then took out home equity loans on the houses to buy more things they could not afford, all because their wages hadn't kept pace with rapidly rising asset prices.

For the most part, the people who issued these mortgages and traded them didn't worry what would happen if the music suddenly stopped. Over the past few decades, the financial services industry in which they worked had grown to represent more than 40 percent of all corporate profits in the United States. Thanks to this "financialization" of America, the soaring stock market and record shareholder profits had very little to do with the actual economy; yet somehow the fate of millions of American citizens—mostly in the form of their mortgages—was tied up in the strange, risky bets of Wall Street banks.

When the music *did* stop, millions of Americans took huge losses. I was one of them. In 2005 my wife and I purchased a home in

Miami. As for many Americans, the purchase was a stretch for us. But financing was readily available, and everyone was led to believe there would never be a housing bubble. Sure, maybe it would happen with tech stocks, but homes had real value. Until they didn't. After the crash, our house was worth less than I'd paid for it. This left us, like many American families, owing more money to the bank than we would be able to get by selling the house. We were in no hurry to move, but if for some reason we needed to, we would have been stuck because it would have cost us money to sell our own home.

We were fortunate; we had the ability to keep making payments and wait out the crisis. Many Americans didn't. In other areas of Florida, people defaulted on their mortgages in droves and left million-dollar homes sitting empty. Within months, there were whole neighborhoods full of half-finished houses and empty swimming pools, all of which had been built to take advantage of the housing boom of the early 2000s. In some cases, speculators—people who had been able to get mortgages with no income verification and were planning on "flipping" the houses for a profit—owned the mortgages on these homes. But more often the people who suffered in the early stages of the financial crisis were ordinary hardworking Americans who had done everything our government had told them they should do. They had taken advantage of historically low interest rates to apply for mortgages, secure in the knowledge that the value of American homes never decreased over time. Unlike cars, which depreciate as soon as you drive them off the lot, homes are supposed to be stable investments. They are known as the lifeblood of the middle class for a reason.

When the financial crisis began, the government could have passed legislation to help these people. Typically, the losses in a financial crisis are shared between creditors (in this case, the banks) and debtors, the people whose mortgages these banks were slicing up and trading among themselves. But the people in charge of negotiating our way out of this crisis prioritized the rights of the banks, leaving ordinary people to absorb most of the pain.

Throughout these negotiations, which occurred during the final months of the Bush White House and into the first part of President Obama's first term, the financial industry that had created this mess effectively held a gun to the head of the American people. The banks insisted that they had become "too big to fail," and that allowing them to go under would tank the American economy.

In a sense, they were right in more ways than they realized. For the past thirty years, in the wake of massive globalization, the American economy had become extremely interconnected with the world, and not always in ways that were beneficial to us. Years later, it would come to light that Russia—a country that was heavily invested in the government-sponsored enterprises Fannie Mae and Freddie Mac—approached the leaders of China, which was also invested, with a plan to sell their shares at the same time, thereby flooding the market and crippling the United States economy even further. It was a dark sign of the conflicts that would soon embroil the world in the years to come.

In the end, the negotiations to save our economy revealed the historic failure of the free-market ideology of Wall Street and technocrats at the Federal Reserve. For years the experts in finance had told us that anything that was good for the market was good for the American people too. They insisted that record profits for shareholders—which had been earned, often, by offshoring jobs and manipulating stock prices—would eventually trickle down to the American people. But it didn't happen. While the average annual household income for the top 5 percent of American earners rose dramatically between 1967 and 2007, the same figure for the bottom half of earners remained stagnant. It's also worth noting that the management teams of major corporations made more money because they were managing more people—it just so happened that now the people they were managing worked in China, not towns and cities in the United States.

In the end, the government granted historic bailouts to Wall Street and did nothing for the millions of Americans who suffered as a result of Wall Street's risky manipulation of American finance. It didn't matter which party was in charge. While the bailouts began under the leadership of Hank Paulson, the former Goldman Sachs CEO who became President Bush's treasury secretary, the government began the process of saving the investment banks, making sure they recouped every dollar of the toxic assets they had lost. Under President Obama, Treasury Secretary Timothy Geithner continued along the same lines.

When the crisis was over, major investment banks had been given billions of dollars to save themselves, while the people whose mortgages they had traded got nothing. These were the same people who would soon lose their jobs in the recession that followed this crisis, and who would struggle for years amid the slowest economic recovery in our nation's history.

There were several lessons we could have learned from this crisis. But we didn't. By 2009, while the unemployment rate was hovering around 10 percent, Wall Street banks regained old strength and then some. Goldman Sachs alone paid out $16.2 billion in bonuses that year, and announced a record profit of $13.4 billion, most of which came from "trading on its own account," in the words of the journalist Andrew Ross Sorkin.[7] Even the most troubled banks soon began trading the same complex, functionally worthless securities that had caused the crisis in the first place, aiding in the continued financialization of the American economy. Those in government—many of whom were veterans of the very financial sector they were supposed to be regulating—allowed this to happen.

Missing from all of this was any examination of the question that should really matter for a country's economy and to its public policymakers—whether the nation is engaged in productive activity. Instead of understanding the financial crisis for the rebuke of what the finance industry had become, policymakers launched another decade of financial excess throughout the most inflationary monetary policy in modern history. Again, this led to record-high stock market figures and historic returns for corporate shareholders. Thanks to low interest rates, more money than ever flowed into risky financial instruments designed to create paper wealth rather than tangible goods.

But investment in the real economy continued to decline. As American businesses spent more and more money jacking up their own stock prices, they spent far less on the research and development that creates new products and jobs. Even as stock prices rose and consumer prices fell, wages remained stagnant and fewer people found reliable, dignified work.

Recently, studies have revealed that corporate managers are under enormous pressure to prioritize short-term profits over long-term strength and to sacrifice the creation of durable value in the pursuit of quarterly earnings. This shift in how we allocate capital has sapped our productive capacity and damaged our ability to provide dignified work.

When dignified work is lost or unattainable, it corrodes the human spirit. Recent years have seen the destruction of jobs that provided a way of life for families and communities for generations. Despite the claims that a "new economy" would rescue them, the modern fabric of American work is not thick enough to sustain them. Entire regions have been hollowed out. Even among those who have succeeded by the terms of the financial economy, there is an inescapable sense that their work is not productive in the way it was for generations prior.

Our failure to prioritize the creation and maintenance of dignified work through investment now presents serious problems. These problems include reduced manufacturing employment, less time for families to have and raise children, population loss in rural America and midsize cities, and lower levels of technological development than rival states like China.

But beyond all these real-world consequences, there is a principle—one that makes sense to any student who has taken basic economics.

In order to thrive, a healthy economy needs to make things, whether it's cars, computers, books, or ball bearings. In years past, when our leaders gave us statistics about the economy, *these* are the things they were counting and reporting on. Today, we talk about dividends and buybacks and complex financial maneuvers that often don't help anyone other than a small number of financiers. During the 2012 presidential campaign, Americans saw that the new standard-bearer for the Republican Party was Mitt Romney, a man whose father had run a car company; yet Governor Romney had gone into finance, which had become America's leading industry.

In decades past, nearly every American was able to locate his or her role in the real economy. Someone milled wheat into flour, someone else turned the flour into snacks, then somebody transported those snacks to the grocery store where other people sold them. The brightest minds in our business community would go to work at this snack food company to improve products. Most importantly, investment bankers were involved primarily in raising capital to expand the business. It was simple, but it worked, and it provided huge opportunity for advancement.

This looks very different from the financialized activity that happens today. In modern times, a private equity firm might see a company with stable cash flows. They will buy this company for a great deal of money, most of it borrowed, and then use those cash flows to pay interest on their debt. Then, at some point, they realize that they can increase cash flows by offshoring jobs. Perhaps an entire factory might move overseas.

At this point, the private equity firm—almost certainly an American company—is not really running another American company. Rather, it is managing a financial asset that represents little more than profits and losses. The productive activity is occurring elsewhere.

When an American company makes something in America, on the other hand, the workers have power. You can't make American cars without American workers, so those workers share in the wealth that is created. If American companies are nothing more than financial assets, then workers have less power; they don't do as well even when the owners of the company make record profits.

My staff and I have attempted to come up with policies that would alleviate these problems. We drew up plans to tax stock buybacks, for instance, and to encourage physical investment in our manufacturing sector. We made plans to expand the federal Child Tax Credit and enact a paid family leave policy.

During my campaign for the White House, these issues were central to my message. After I dropped out of the race, it became clearer than ever that Americans were worried about the widening gap that existed between the wealthiest Americans and themselves. The two candidates who tapped into this anxiety—Bernie Sanders on the left and Donald Trump on the right—did remarkably well during the primaries. Hillary Clinton, who was seen as being too cozy with Wall Street, did not, only winning the nomination after the Democrat establishment conspired to knock Bernie Sanders off the top of the ticket.

For the next few years, I continued to push these policies in the Senate, often encountering resistance from both sides of the aisle. From the left, I'd hear lame excuses as to why we couldn't get tough on China. Maybe they truly believed Chinese solar panels and batteries would save the world from a climate catastrophe, or maybe it was just that their corporate donors made a bunch of money there. And ironically, from the right, I'd get a lecture about why we had to defend corporations from the radical left even as the left was weaponizing those corporations against conservatives.

Common-Good Capitalism

Today, the debate around financialization is not an easy one to have. The issue does not fit neatly into our standard partisan framework, which has been eroding for years. While the left often attempts to paint the right as rich and out of touch with working people, I have found during my travels around the United States that the opposite is usually the case. There is a reason that Hillary Clinton, the Democrats' preferred candidate in 2016, was not elected to the presidency. The American people have been waiting too long for someone to address the problems that exist in our economy, and they are tired of watching as candidates come into office and fail to take significant action to address the issues they campaigned on.

The answer, I believe, should begin with a national consensus that our challenge is not simply one of cyclical downturns or having the wrong party in charge. As we've seen for the past few pages, this problem comes from both sides of the political aisle. Our challenge is an economic order that is bad for America—one that is bad because it is leaving too many people behind. Agreeing on the problem is something we should be able to achieve across the political spectrum.

Deciding what government should do about it should be the core question of our national politics. The old ways simply will not do. The notion that the market, left unguided and unchecked, will solve our problems will not restore a balance between the obligations and rights of the private sector and working Americans. It may lead to GDP growth and record profits—at least for a while—but economic growth and record profits will not, on their own, lead to the creation of dignified work.

Of course, socialism would be even worse. The idea that government can single-handedly impose a balance between the obligations and rights of the private sector and working Americans has never worked. We have millions of Americans who came here to flee socialism who can testify to that. They know that a government that guarantees you a basic income is also one that controls where you work and how much you make. They know that a government that guarantees you free health care is also one that controls who your doctor is and what care you'll receive, and that a government that promises you free college is also one that controls what school you go to and what you are taught.

What we need to do is restore what I call common-good capitalism. This is a recognition that, first, capitalism is the greatest economic system that has ever been created. It allocates resources with a level of efficiency that cannot be replicated anywhere else. But until recently, very few people thought that capitalism was perfect. Fifty years ago, nobody worshipped at the altar of the free market. Irving Kristol—one of the great conservative intellectuals in American history, for whom the American Enterprise Institute names its highest honor bestowed upon conservative leaders—wrote a book called *Two Cheers for Capitalism*. It was two cheers, not three, because he understood that capitalism isn't perfect and that capitalists need to care about the health and values of a nation, not just their bottom-line profits.

I think of capitalism a lot like I think of fire. Human civilization couldn't exist without fire. Fire is an important and good thing. But it is not an absolute good. Fire run amok is destructive. Fire must be tended to and cared for. The same is true for capitalism. Capitalism must be cared for by capitalists who prioritize the well-being of their workers, the strength of their communities, and the health of their nation. Capitalism must be cared for by workers who understand their obligations to work and to contribute responsibly to our nation.

Capitalism also needs to be cared for by policymakers. Our economy is supposed to work for our nation; our nation does not work for our economy. And while capitalism will always allocate resources efficiently, the most efficient allocation of capital is not always what is best for our nation. It is not best for our nation if we are reliant on China for our most vital pharmaceuticals. It is not best for our nation if our national freight rail lines are so efficient that a rail worker has to show up for work even when he is extremely sick. It is not best for our nation if the demands of the economy are so great that a new mother does not have the option, if she wishes, to stay home and care for her child.

I think of my role as a policymaker as very similar to the role of the commissioner of the National Football League. In football, the goal of a team is to compete as ruthlessly as possible to win as many games as they can by as many points as possible. The head coach of the Miami Dolphins would be very happy winning every football game 49–0. No coach ends up in the hall of fame for consistently going 9–8. Close games only create the possibility of a bad bounce costing you a win. But even as big a Dolphins fan as I am, I would not want them to win every game 49–0. It would be fun once or twice, but it would be boring to watch all season long and season after season.

The role of the NFL commissioner is to make sure that the rules of the sport channel the competitive impulses of each team to create an outcome that the fans want—entertainment. When the game got boring because there wasn't enough scoring and passing, the rules were changed to protect the passer. Nowadays, you cannot hit the quarterback high, you cannot hit the quarterback low, and you cannot hit the quarterback late. The result is a tremendous offensive boom in football and a sport that is more fun to watch than ever. The rules were created to incentivize the teams to do what the sport needed.

That's what I mean by common-good capitalism. It is a form of capitalism that is tended to by stakeholders to ensure that the results our economy produces are in the best interest of our people, our families, our communities, and our nation.

For example, our tax code is still biased in favor of stock buybacks. I'm not saying that these should be illegal. But they do not boost job creation or worker pay, so it stands to reason that they should not have a tax preference. Instead, tax preferences should be for the use of corporate profits that further the common good by creating new jobs or higher wages. This is why we should make immediate expensing a permanent feature of our tax code; we should give a tax preference to businesses when they reinvest their profits in a way that creates new jobs and higher paychecks.

Common-good capitalism also means recognizing that the market may determine that outsourcing industries like manufacturing is the most efficient use of capital. But our national interest and the common good are threatened by the loss of these industries and capacities. For two decades, we have allowed competitors like China to use subsidies and protectionism to build up their capabilities in various key industries while destroying ours. For example, rare earth minerals are vital to our national security because they are a critical component of specialized computer and weapons systems. Mining them is a source of stable and dignified work. But we have allowed America to become almost completely dependent on China for rare earth minerals, and we have done nothing to further our ability to provide them for ourselves.

There are many emerging industries to which we should take a similar approach—where promoting the common good will require public policies that drive investments in key industries, because pure market principles and our national interests are not aligned. Aerospace, telecommunications, autonomous vehicles, energy, transportation, and housing are just a few of the industries in which America must always retain not just domestic capacity but also global leadership.

The goal isn't to re-create the economy of 1969. That's impossible, and it would be foolish to try. Many of the advances that have been made during the era of financialization *do,* in fact, help American workers. But too many do not. Learning the difference between the two will be difficult, but it is necessary if we are to rebalance our economy and rebuild a nation that is capable of delivering the good life for anyone who is willing to put in the work and be part of this incredible nation we call home.

Chapter 3

THE RISE OF CHINA

Fueling China's Rise

History is filled with examples of one power rising at the expense of another. China's continued growth is inevitable, and not something America should try to stop. It is already a large and powerful global power, and it will continue to be one. But there is a big difference between a large and powerful China and a large and powerful China that tries to rise at our expense. Now, free-market fundamentalists and Wall Street traders hate the phrase "rise at our expense." As do executives in corporate boardrooms and so-called experts at think tanks. No one has risen at our expense, they say. Life has never been better in America, they argue.

But a nation isn't an economy. It is a group of people living together in community. And when you look at the people and their communities, the stats look very different. Income inequality is up. Homeownership is down. Overdoses are up. Birth rates are down. Crime is up. Good jobs are down.

This is what it looks like for China to rise at our expense.

When our country allowed China to join the World Trade Organization in 2001, we kicked off a chain of events that would inevitably end with the rise of a near-peer competitor. The idea, which despite the obvious warning signs I discussed earlier was supported by just about everyone in this country with an Ivy League degree, was never seriously challenged by those in a position to make a difference. Those on Wall Street said it would boost profits for American corporations, which it did. We allowed ourselves to believe that our material desires and global stability were perfectly aligned because commerce would lead to peace.

We are reaping the disastrous consequences of that foolish belief.

Even worse, major figures in the United States continue to fuel China's ascent toward global supremacy. Even after the events of the past thirty years, these people continue to believe that the government of China will liberalize and move toward greater openness and democratic values. Either that, of course, or they *pretend* to believe these things in public while knowing it's all hogwash. "Who cares if America is in decline?" they might tell themselves. "I can't do anything to stop it and I'm getting fantastically rich in the process." I'm sure not all of them are that cynical, but the results are the same either way.

American politicians, businesspeople, and celebrities pump trillions of dollars into China's economy every year, enriching corporations that are explicitly tied to the Chinese Communist Party. When an American citizen does business with a bank in China, for instance, that person is effectively doing business with the Chinese Communist Party. Over there, the difference is minimal at best. When Hunter Biden, for instance, began negotiating with the energy tycoon Ye Jianming, American news sources pointed out that "at its height, Ye's company, CEFC China Energy, aligned itself so closely with the Chinese government that it was often hard to distinguish between the two."

Speaking to a group of his compatriots during the Russian Revolution, the Marxist thinker Vladimir Lenin is rumored to have said that "the capitalists will sell us the rope with which to hang them." Every day in the United States, elite financiers and politicians give China more and more rope; it is only a matter of time until they begin stringing that rope around our necks. It should come as no surprise, by the way, that General Secretary Xi Jinping counts Vladimir Lenin among his great heroes and exhorts the Communist Party to "liberate all of humanity" and serve as the "gravediggers of capitalism." American elites may be done with the dialectic, but the dialectic is not done with them.

As of this writing, the list of Americans fueling the rise of China is long. It includes the managers of major hedge funds that have a great deal of money invested in China, and whose partners have developed close ties with officials in the Chinese Communist Party. This is all part of the strategy, as outlined by Professor Anne-Marie Brady, to "forge close partnerships of mutual advantage."

But there is perhaps no greater booster of China in the United States than Ray Dalio, the billionaire founder of Bridgewater Capital. Other than Bridgewater's business in China—which includes investments of millions of dollars—Dalio has written in glowing terms about China's repressive government, euphemistically calling it "paternal." He argues that in China, the government believes "a steady hand is needed to maintain order." He also quotes the government as saying that "what happens inside [China's] borders is its business, and the U.S. has its own human rights problems," and does not disagree in the slightest with either of the claims. In the past, as the journalist Peter Schweizer has noted in his book *Red Handed*, Dalio has said that he will not judge which system of government—America's or China's—is "better."[8]

For years, Dalio spoke well of Jack Ma, the Chinese entrepreneur who had been very mildly, but still openly, critical of some regulations put forward by the CCP. Then, when Ma "disappeared" as a result of his battle with Chinese regulators, Dalio took the side of the government. He called Chinese regulators "reasonable, caring, and highly informed."

He went on to say that people have accused him of being "biased, naive, and unpatriotic," but he thought he was just being "objective."

Dalio's comments may seem jarring, but at least he's transparent about his motivations. He sees America in decline and hopes to profit from it. That is grotesque, but at least honest. National policy debates would be much better if everybody was as honest about their motivations.

For example, when companies like Nike, Apple, and Coke lobbied against my bill to crack down on Chinese imports made with slave labor, they typically talked about how the law would impact American consumers and thus American jobs. What they meant but couldn't say is that cracking down on slave labor in China would hurt their profit margins. They didn't want to make things in America. Heck, once Chinese labor became available, these companies didn't even want to make things in our hemisphere anymore. They didn't care about American communities five or ten years down the road. They cared about their next quarterly profit report.

The same goes for all these people talking about solar panels and electric vehicles. China dominates those sectors, in terms of both controlling the natural resources and the prevalence of slave labor that keeps costs low. If Tesla cared about America's future, it would prioritize expanding production lines in the United States, not opening a showroom in Xinjiang or new factories in Shanghai.

Many Americans, especially those in business, finance, and politics, cannot imagine a world in which China and the United States are not tied at the hip. We've become so intertwined, so reliant, that decoupling seems too difficult to even contemplate. Even in critical areas like semiconductors, advanced biomedical research, and rare earth minerals, the status quo dictates that we remain coupled with a Communist regime determined to inflict a century of humiliation on America.

Again, we did not get here by accident. Our reliance on China—financial and otherwise—is the direct result of decades of bad US policy. Beyond the story of how we helped China to enter the WTO, there is a much broader story about how we willfully misunderstood the country's ambitions from the beginning, and how doing so allowed the relationship between our two countries to deteriorate into a state resembling open war.

Blind Ambition

With the benefit of hindsight, it is easy to see China's single-minded determination to rise to the top of the global order. Many of the most adamant proponents of integrating China into the global economic system will now acknowledge the Communist nation's raw ambition. Some even understand that their assumptions two decades ago were completely wrong—at least those about how the Chinese Communist Party would "modernize" and "evolve."

Perhaps our leaders were simply naive—others would be less charitable. But the end result is the same. China exploited America's blind ambition—ambition to open markets, ambition to spread democracy, ambition to make obscene amounts of money—and we are now paying the price.

In some sense, the story begins in 1974. That year, a congressman named George Bush was named President Nixon's unofficial ambassador to China. During his time in Beijing, Bush became close with several high-profile members of the Chinese Communist Party. Just before he left, Deng Xiaoping, then the vice premier of China and soon to be paramount leader of the People's Republic of China (PRC), said they were "old friends," encouraging him to "come back anytime."

Nearly two decades later, after Bush was elected president, he kept close ties with Chinese leaders and insisted on personally overseeing relations with the country. During his time in President Bush's cabinet, Secretary of State James Baker would often refer to his boss as "the desk officer on China." Like many people who served in government at the time, President Bush believed that the world was getting simpler. After the fall of the Berlin Wall, it was widely accepted that the threat of communism was gone, and that the only workable form of government left was liberal democracy. Unfettered capitalism would transcend political divisions, blur nation-state borders, and usher in an era of unrivaled peace and prosperity.

The State Department, led primarily by James Baker and his deputy Robert Zoellick—both career public servants who'd come up through the Ivy League and gone straight into government—began crafting policy on the assumption that all the United States had to do from now on was spread liberal democracy and free-market economics.

Within the next few years, the Bush administration decided that it wouldn't be long before every country in the world followed suit. In their eyes, the struggle to find the best form of government was over. So was the era of air raid drills, espionage, and great power politics that had characterized our struggle against the Soviet Union. Moving forward, they believed that the world, in the words of Thomas Friedman, was about to become "flat." Companies would expand all over the globe and give more countries than ever access to the global market. In turn, those countries would come to look like liberal, free-market versions of the United States. The few nations left in the world that were still openly Communist—my parents' home country of Cuba, for instance, or the small, impoverished totalitarian kingdom of North Korea—would either learn to adapt to this new reality or sink even further into obscurity.

Here again we see American leaders—respected public servants well versed in law, history, and our Constitution—making a simple but catastrophic mistake. Almost to a person, they believed our uniquely American values were downstream from the economic system developed by America over the previous century. Capitalism can exist and even flourish without protecting our God-given rights. Without values, capitalism, markets, and finance are simply tools to be used by whoever has power. Giving authoritarians more power is rarely a way to bring about peace.

In June 1989 China's government murdered hundreds, if not thousands, of Chinese students who were protesting *for* democracy in Tiananmen Square. If anything, this event should have convinced our leaders that we should be skeptical of China's promises as it was stepping onto the world stage. It should have been a wake-up call for our nation's leaders, an excuse to begin taking slow, deliberate steps rather than charging full steam ahead.

But that's not what we did. Despite some outcry from lawmakers on both sides of the aisle in Congress, President Bush wrote a letter to Deng Xiaoping and sent Brent Scowcroft, his national security advisor, as a secret envoy to keep the friendship between the two countries alive. Days later, after the massacre was complete and the bodies were being carried away from Tiananmen Square, President Bush held a press conference in the White House, during which he claimed that "the forces of democracy" in China were now so powerful that they would be able to "overcome these unfortunate events." Later he said that the rise of liberalism in China seemed so inevitable that it would be impossible to "put the genie back in the bottle."

Six months after the massacre, Brent Scowcroft returned to Beijing—this time publicly—and had dinner with his Chinese counterparts. During a candlelight dinner they made a champagne toast, "as friends, to resume our important dialogue."

For years the Bush administration crafted policy based on these wrongheaded assumptions. Although they did ultimately pass sanctions against the Chinese Communist Party for the Tiananmen Square massacre, those sanctions were widely viewed as being far too lenient. Writing about the event nearly a decade later in his memoir *A World Transformed,* President Bush attempted to explain his thinking.

"Tiananmen shattered much of the goodwill China had earned in the west," he wrote. "To many it appeared that reform was merely a sham, and that China was still the dictatorship it had always been. I believed otherwise. Based on what I had seen over the previous fourteen years, I thought China was slowly changing and that the forces of reform that had been building were still strong."

Two years after the massacre, President Bush delivered a version of this message to Yale University's graduating class of 1991. Speaking from the campus in New Haven, Connecticut, that had produced many of our nation's presidents and foreign policy elites, he said that "the most compelling reason to . . . remain engaged in China is not economic, it's not strategic, but moral. It is right to export the ideas of freedom and democracy to China."

He continued: "If we pursue a policy that cultivates contacts with the Chinese people, promotes commerce to our benefit, we can help create a climate of democratic change. No nation on earth has discovered a way to import the world's goods and services while stopping foreign ideas at the border. Just as the democratic idea has transformed nations on every continent, so, too, change will inevitably come to China."

The naïveté and arrogance were stunning, even then. The CCP was telling the world in no uncertain terms that it was operating based on a different set of ideals than those on which we pretended the world operated, but even a mass slaughter of civilians wouldn't cause us to abandon our ideological conviction that commercial contact would lead to democratic opening. No two nations with a McDonald's would ever go to war—it was written right there on the ten commandments of neoliberalism!

Blind ideological faith, however, was not prudent public policy. There's a good reason that when cabin pressure on an airplane is decreasing, the flight attendants will tell you to adjust your own oxygen mask before helping others; they know that if you've passed out, there's very little you can do for anyone else. Along the same lines, leaders in the United States should have paid more attention to our own ability to make things before turning Communist China into a global superpower. Instead, we proceeded from the naive belief that we were invincible—that no matter how powerful China became, we would remain powerful. We did not tighten our own oxygen masks first. We assumed we'd never need one again.

It mattered little which party was in the White House, or which party was in control of Congress. Somehow, the notion that we should welcome all countries into the global order managed to become gospel with almost no debate, resistance, or analysis of the risks. Even when members of Congress did raise concerns, their votes mattered so little that the arguments faded away quickly.

China's rise gained momentum under President Bill Clinton. Like his predecessor, he had a government full of people telling him that bringing China into the global market could only have good consequences. Anyone who said otherwise was dismissed.

Despite having attacked his predecessor several times on the campaign trail for being soft on "the butchers in Beijing," the newly elected president soon abandoned that line of attack, largely due to pressure from elites in Washington think tanks and on Wall Street who believed that China's human rights abuses were a small price to pay for soaring corporate profits. Again, we were blind to the horrific human rights abuses that occurred every day in China.

But now that blindness was more willful than ever.

In 1993, for instance, President Clinton signed an executive order demanding that Beijing make "significant progress" on human rights in order to retain its trade privileges with the United States. At the time, China was not yet a member of the still-nonexistent World Trade Organization, so its most-favored-nation status was up for review every year. In response, officials in Beijing rejected his demands. One year later, Clinton "reversed course and restored China's status without condition of threat of further review."

In 1994 Clinton moved to "delink" human rights concerns from the annual most-favored-nation extension process altogether, meaning that the United States could not consider the country's vast array of human rights abuses when deciding whether to make new trade agreements. That same year, over the strident objections of labor unions and human rights groups, Congress voted to uphold his decision.

All that mattered, it seemed, was that the market stay open.

There were signs, of course, that Americans didn't go along with the elite consensus. Patrick Buchanan's shocking victory in the New Hampshire primary in 1996 caused more establishment heartburn than the Tiananmen Square massacre seven years earlier. Buchanan, of course, famously suggested an embargo of Chinese goods, calling the current approach a "reflexive accommodation and appeasement of Communist China."

During a speech that he gave at the White House shortly before China joined the WTO, President Clinton briefly considered the devastating effect that the internet—still relatively new at the time—might have on the Chinese Communist Party's ability to censor and oppress its citizens. By this time it was common knowledge that they were attempting to do so, and several critics of the administration's soft China policy had raised the issue. Again, President Clinton was dismissive.

"Now there's no question China has been trying to crack down on the internet," he said. "Good luck! That's sort of like trying to nail Jell-O to the wall. But I would argue to you that their effort to do that just proves how real these changes are and how much they threaten the status quo. It's not an argument for slowing down the effort to bring China into the world, it's an argument for accelerating that effort. In the knowledge economy, economic innovation and political empowerment, whether anyone likes it or not, will inevitably go hand in hand."

As we've already seen, these two things do *not* necessarily go hand in hand. It's perfectly possible for a poor authoritarian state to take in trillions of dollars and then use those trillions of dollars to get more authoritarian, not less. When a drug dealer makes millions of dollars conducting his business, to use an extreme analogy, he doesn't give up dealing drugs, become an accountant, and start donating to charity every year. Instead, he uses that money to keep doing what he was doing in the first place—more dealing, more crime, and exponentially more money from all of it. And suddenly, whether through money-laundering operations or legitimate business investments, the drug dealer is now supporting a significant amount of economic activity.

During the Clinton administration, the same thing happened to China. Rather than getting freer and more open, they got less free and less open; but the more money they were making, the more industries, corporations, and investment funds became entangled with the Communist regime. Just like with the drug dealer, it was easier to ignore the character of the government and focus on the financial rewards. After all, what you can't see can't hurt you.

Again, the trends were there for anyone who wished to see them. In 2002, for instance, after thirteen years of rule by Jiang Zemin, another leader of the Chinese Communist Party emerged. This one, Hu Jintao, put on an open face for the global economy. China, he insisted, was ready to adopt free-market principles and serve as a manufacturing base for major American companies. But internally, the repression of the CCP's enemies grew worse than it had ever been.

Hu came to power largely because he had demonstrated an ability to harshly repress dissent in the Tibet Autonomous Region of China, which he oversaw on behalf of the CCP. But to the world, he talked about "China's peaceful rise," and generally pushed foreign policies that were friendly to American neoliberals who wanted to engage with China.

It worked.

Speaking shortly after he was elected, the second President Bush spoke about China in the same glowing terms that his father had used a decade earlier in his speech to the students of Yale University: "The case for trade is not just monetary but moral. Economic freedom creates habits of liberty. And habits of liberty create expectations of democracy. . . . Trade freely with China, and time is on our side." It should come as no surprise that the son echoed the father despite another decade of evidence. All the same players were back in power, and they would be damned if they were about to change their strategy.

A few years later Robert Zoellick—who joined the younger President Bush's government to further the mission that had begun under his father—extended this reasoning. Now that Beijing had become "a player at the table," the Communist Party had an obligation to become "a responsible stakeholder," helping to "strengthen the system that enabled its success" by increasing transparency about military activities, addressing its human rights abuses, and protecting the intellectual property of companies that do business in China. Repeatedly, Zoellick was one of the strongest voices for bringing China into the global economy, often speaking about the wonderful effects that greater cooperation would have.

But none of it happened.

By the time Barack Obama insulted half of America as being bitter racists clinging to their guns and religion, the Chinese Communist Party was already ascendant. As America began to crack from within, the rest of the world realized that the "end of history" had been a naive, half-baked thought all along. The professors, Wall Street bankers, and op-ed writers who'd pushed the idea in the first place issued their usual weak apologies in newspapers and academic journals, tossing the idea aside as they had so many others before it.

Writing in *Foreign Affairs* in the spring of 2018, two analysts named Kurt Campbell and Ely Ratner—both of whom are now senior officials in the Biden administration—admitted that the "policies built on such expectations have failed to change China in the ways we intended or hoped," but also claimed that no one could have foreseen the various ways things had gone wrong in the past thirty years. This is demonstrably false, as any careful study of Chinese history during the period would reveal.

But their attempts at revisionism hardly matter now.

By the time our elites opened their eyes and saw China for the rising authoritarian power that it had become, it was too late, as President George H. W. Bush had predicted in the aftermath of Tiananmen Square, to put the genie back in the bottle.

It just wasn't the genie he'd been expecting.

In 2012 the Chinese Communist Party installed Xi Jinping, a fiercely nationalistic leader and fervent Marxist-Leninist who abandoned all pretense of a "peaceful rise" for China. Under Xi, the

Chinese Communist Party has begun calling back more than ever to the era of Mao Zedong. After five years at the head of the CCP, Xi wrote his own name into the Chinese Constitution in the form of the phrase "Xi Jinping Thought on Socialism with Chinese Characteristics for a New Era." In doing so, Xi has attempted to place himself among the ranks of famous Chinese leaders such as Mao and Deng Xiaoping, and also to cement his economic and governance philosophy into Chinese life forever.

Domestically, Xi has been more totalitarian than any leader of the CCP since Mao. When the coronavirus broke out in Wuhan, he had no qualms about locking people in their homes to deal with the virus. Under Xi's draconian "zero-Covid policy," countless Chinese people have suffered and been dragged away by the authorities for fear that they might infect their neighbors with Covid. Others have been forced to stay home in horrible conditions for months at a time.

As we've already seen, Xi has also cracked down on the Uyghur Muslim population in Xinjiang, sending members of that ethnic group to slave labor camps where they've endured systematic rape, forced abortions, and medical experiments. But none of it seems to matter to corporate leaders. The multinational consulting group McKinsey & Company should never have hosted a corporate retreat only a mile away from one such concentration camp. Disney even thanked the regional security forces for help in filming *Mulan*.

When it comes to power, Xi has shown that he has no qualms about doing what is necessary to maintain his grip on the party. A few years ago, he "disappeared" the famous entrepreneur Jack Ma—a member of the Chinese Communist Party—for two months as a show of strength to the technology sector.

In terms of foreign policy, Xi has been explicit about confronting the United States, and has aggressively expanded China's territorial claims in the South China Sea, essentially turning tiny drifts of sand into massive military complexes.

But that is only the beginning.

A Great Wall of Steel

China is not a liberal democracy. It is a totalitarian Communist regime whose rise is the product of market-distorting policies, and the CCP has no plans to ever change that. For them, industrial and trade policy isn't about improving the well-being of the Chinese people. It is about increasing their power relative to ours. This is a goal they have pursued with ruthless efficiency.

Under the leadership of General Secretary Xi Jinping, the Chinese Communist Party has emerged as the most formidable economic and military opponent that the United States has seen since World War II. And unlike the United States, China's ambition is fueled not by a desire for global freedom and prosperity but by a desire for complete and total global supremacy. In the eyes of China's Communist leaders, China's history is one of thousands of years of glory that have only recently been interrupted by Western meddling.

Most troublingly, some of the top supporters of the Chinese Communist Party are citizens of the United States who have been investing in China's rise for years, waiting for the inevitable rise of this once-backward economy. Many of these people currently work on Wall Street, where billions of dollars of American capital are pushed into Chinese investments every day. Others are the founders and current leaders of tech companies in Silicon Valley who see China as an ally in manufacturing and an enormous market for their products.

Too often, these people are former government officials—the same ones who made the policies that pushed the United States into the arms of China in the first place. After assuring us that Beijing would become "a responsible stakeholder" in the global economy, for instance, Robert Zoellick went straight to Wall Street, where he enjoyed a cozy relationship with the Communist leadership as a senior international advisor to Goldman Sachs.

Despite ample and widely available evidence that the Chinese

Communist Party has become a brutal totalitarian state bent on world domination, many Americans will defend and defer to them whenever they feel it is necessary. Examples of such fealty are abundant, especially in the last few years. When the actor John Cena referred to Taiwan as a country in a speech, he issued an apology—in terrible Mandarin Chinese—for the offense. When Houston Rockets general manager Daryl Morey tweeted in support of protestors in Hong Kong, LeBron James apologized for him, saying he wasn't "educated on the situation."

In 2012 former British prime minister David Cameron met with the Dalai Lama in London. Almost immediately the Chinese Communist Party, which was in the midst of a very public disagreement with the Dalai Lama over Tibet, demanded an apology. The party called off investment in Britain for a while and canceled several diplomatic meetings between the two countries.

Eventually, the British government issued a formal apology. According to an account of the meeting later given to a journalist by someone who was present, "Before the meeting had got underway the CCP officials pushed a copy of the British apology across the table toward their British counterparts, who were then asked to stand up and read it aloud, which they duly did. Sitting down afterward, the lead Chinese official reportedly smiled and said, 'We just wanted to know you meant it.' "

Clearly, the party is no longer hiding its plans for global supremacy.

In the summer of 2021, for instance, the Chinese Communist Party celebrated its one hundredth anniversary. Speaking from Tiananmen Square, General Secretary Xi spoke of a party that had risen from humble beginnings to become the most powerful political force in the world. The Chinese Communist Party, he said, formed in 1921, now boasted more than 95 million members. Their country, once an economic backwater, now had the second-largest GDP in the world.

"Through tenacious struggle," Xi said, "the Party and the Chinese people have shown the world that the Chinese nation has achieved the tremendous transformation from standing up and growing prosperous to becoming strong, and that China's national rejuvenation has become a historical inevitability."

For the next few minutes, Xi quoted Chairman Mao Zedong and praised his colleagues in the Communist Party. Then he turned his attention to the wider world, particularly the West.

"We will never allow any foreign force to bully, oppress, or subjugate us," he said. "Anyone who dares to try will find their heads bashed bloody against a great wall of steel forged by over 1.4 billion Chinese people."

Hours later, when the party released its official English transcript of this speech, the words "bashed bloody" were removed. But the general tone of hostility toward Western countries remained.

Reading this speech, it's hard not to marvel at how we came to live in a world where China—a country that was once at the bottom of the global economic ladder—is able to intimidate the world with language about "heads bashed bloody against a great wall of steel." Indeed, the rise of China was not "a historical inevitability," but the result of a series of extremely bad decisions, almost all of which were made by the elite political class in the United States. As China rose, officials from the US government helped them at every step of the way, failing all the while to see the threat that was right before their eyes.

Historical precedents for such a blunder are difficult to find. But they do exist. In the aftermath of the Vietnam War, for instance, when it was becoming clearer every day that our involvement in the conflict was a blunder of historic proportions, the journalist David Halberstam wrote a book titled *The Best and the Brightest*. His mission was to find out exactly how a group of young men who had all come from the best families, gone to the best schools, and prepared themselves for government service in all the traditional ways had managed to lead this country into the worst, least winnable conflict of the twentieth century.

Despite having the right degrees and a stellar ability to absorb the latest academic ideas of their time, the men who served in the cabinet of President John F. Kennedy (and later Lyndon Johnson) failed spectacularly when it mattered most. As a result, nearly sixty thousand American soldiers were killed, and America's standing on the world stage took a major hit—one from which it has yet, in some sense, to recover fully. The question that Halberstam spends around five hundred pages investigating is *how* they managed to keep making the same bad decisions over and over again, even as the horrid consequences of their actions showed up daily on the front pages of newspapers.

The answer, which comes near the end of the book, is simple. These men "had, for all their brilliance and hubris and sense of themselves, been unwilling to look and learn from the past." As Halberstam writes, "They had been swept forward by their belief in the importance of anti-communism (and the dangers of not paying sufficient homage to it) and by the sense of power and glory, omnipotence and omniscience of America in this century."

A little over two decades later, a new group of bright, well-educated young men and women emerged to call the shots for the United States. They were filled with the same "hubris," and the same "sense of power and glory, omnipotence and omniscience of America in this century" that had plagued the men who had led this country into the Vietnam War and kept us there for far too long.

But today, the fad ideology of the age is not anticommunism. Today, neoliberalism is in. In the eyes of our elites, the spread and support of free trade should come before all other concerns—personal, practical, and geopolitical. In recent years, this has led to a kind of "free-market fundamentalism" that is, as the journalist Joel Kotkin has put it recently, "as ideologically brittle, entrenched, and impervious to critique as any Leftist vision of social utopia."

If we had approached Beijing with even a touch of skepticism about this free trade, no-questions-asked mindset, we would almost certainly be in much better shape domestically. China would likely still be a rising power. It is a massive nation, with a huge population, a wealth of natural resources, and a long history of innovation. The question was always how it would rise, not whether it would. Bush and Clinton had that part right, but that was the only part they had right. As it turns out, accelerating the rise of a brutal Communist regime turned out poorly for America, the region, and, of course, anyone who dares voice dissent in China. And while it is impossible to prove, I have no doubt China would not be nearly as powerful as it has become without the unprecedented support we've given its government over the past three decades.

Surely, some in the ivory tower and in our media would be telling us that we missed an opportunity to bring China into the global economy and help the country liberalize its government. They would say that America had a unique historical opportunity to turn Beijing toward us, but instead we refused to extend an open hand, and the PRC went in the wrong direction as a result. For three decades, we have tried governing our country their way, and we have a catastrophe on our hands. There is still time to stop listening to them and do something else.

For the last thirty years, our policy toward China has not changed their behavior. In fact, we explicitly avoided addressing their behavior, assuming the market would fix it for us. Could we have changed behavior if we'd been less blind to China's ambition and our own? Perhaps, but one thing is for sure: China's rise would have been much slower had we not bent over backward to accommodate Beijing and Wall Street. And we would be much stronger and capable of managing our relationship with China.

When it comes to America's policy toward the People's Republic of China, there has been nearly unanimous agreement among our elites on one specific policy approach. With regards to his administration's policy toward the Soviet Union, Ronald Reagan used to talk about "trust, but verify." This would have been a wise approach to the rise of the People's Republic of China. It has always been the case, and still will be the case, that China will become a powerful and influential nation on the world stage. It is simply impossible for a nation the size of China and with its rich history to be anything less. But how it gets there is the critical question. If China seeks to rise in a way that respects other great powers and works within the largely peaceful and prosperous framework that has served the world well for decades, then there would be no reason for concern. This, however, is not the path the Chinese Communist Party is taking. Rather, its actions indicate it seeks to rise at the expense of other powers—and, specifically, by supplanting the United States as the most influential nation in the world.

Our nation hasn't taken a "trust, but verify" approach with China over the past thirty years. Instead we adopted a nearly religious conviction that if we made China richer, it would integrate peacefully into the world order. That wealth and consumption would make it impossible for a nation to harbor any ambition beyond getting more wealthy and consuming more. Depressingly, it seems we projected some of our own weaknesses onto the other. The result of this zealous conviction in the gospel of prosperity is that our government decided repeatedly to fuel China's rise at the expense of our nation's industrial capacity, our national security, and the strength of American workers.

Again, it is not too late to change this. There are immediate steps we can take to ensure that the damage stops immediately.

Chapter 4

CRISIS OF CONFIDENCE

Balance of Power

At the beginning of this book, I suggested that the end of the Cold War was the moment in which a misguided notion began to take hold of our elite political decision-makers. The idea, broadly speaking, was that we had reached the end of history—that after many centuries of grand conflict, the world was finally settling into a peaceful era. Borders would come down, and a new global order would emerge. This stable order would be enforced by the United States of America, which was, at the time, the last true remaining superpower in the world.

There seemed to be good reason to believe this in the early 1990s. The struggle that had defined American life for much of the twentieth century had finally ended, and the United States had won. Through sheer force of will, it seemed, this country had been able to defeat the Soviet Union, one of the strongest and most sinister forces America had ever faced. After that, it felt like there was nothing that we could not accomplish. Not since the end of World War II had there been such a strong belief that the United States could do almost anything it wanted to on the world stage.

This idea was only reinforced in the summer of 1991, when the United States military stopped Saddam Hussein from invading the neighboring country of Kuwait. The ground campaign lasted all of one hundred hours. Shortly after the successful military operations were completed, President George H. W. Bush declared that the Gulf War—the first war that the United States had carried out since the end of the Cold War—had proven that the United States military was a force to be reckoned with. The success of this invasion and its short timeline instilled in our foreign policy elites a belief that we had unrivaled power that should be used as a force for global good.

In part, that was true. With Communist states crumbling all around the globe, the United States was the only country left with the ability to stage interventions in the affairs of other nations. We were certainly the only country left that could do so with any degree of success. That is why our leaders and major thinkers spent so much time over the course of the next few years deciding how and when it was appropriate to use this newfound power to influence affairs around the globe. It seemed that every few years, another crisis required our attention.

In the early 1990s, for instance, American citizens began reading reports from the Balkans, where the Communist government of Yugoslavia had collapsed just a few years earlier. Since the end of the Cold War, the six independent countries that had once been united under that government devolved into various civil wars and ethnic conflicts. Soon, the United Nations declared that some parties in these wars had engaged in genocide, the first time they had used the term since it was applied to the crimes of Adolf Hitler and the Nazis after World War II. Many people in our government declared that the United States had no choice but to intervene, citing promises we had made to stop genocide wherever it occurred in the world. In the end, the United States intervened along with NATO forces and brought about relative peace in the area.

Other cases were more complicated. In one presidential debate in the 2000 election season, both Al Gore and George W. Bush were asked about eight presidential decisions to send American troops into foreign theaters. Both opposed two out of the eight interventions. For his part, Bush made it clear that "our military is meant to fight and win wars. . . . And when it gets overextended, morale drops." Looking back, it was a prescient warning; exactly eleven months later came the terrorist attacks of September 11, 2001. It forced policymakers in the United States to confront extremely difficult questions about our place in the world. Nearly everybody agreed we needed to bring the terrorists who had attacked us to justice and to make the Taliban regime that harbored them pay a price, for instance, but what should the goal be when that invasion was completed? How much could America really do to create a world order in which further attacks of such magnitude would no longer be possible?

Those were important and relevant debates. Without playing Monday-morning quarterback, there is one thing we can say for sure: America—under Republican and Democrat leadership—suffered from a mix of totally justified rage and hopelessly naive optimism, both of which blinded us as the global order shifted beneath our feet. While the United States spent twenty years focused primarily on the Middle East and counterterrorism, China was growing stronger—economically, diplomatically, and militarily. America led a unipolar world after the end of the Cold War. That world still existed in 2001. But America could not remain the world's sole superpower forever. Eventually, it should have been clear that a moment would come when a rising power—China, for instance, probably in an alliance with several other nations and groups hostile to the interests of the United States—would attempt to supplant our nation as the keeper of the global order.

One thing is the same today as it was 5,500 years ago, and will be the same 5,500 years from now: human nature. Human nature, driven by the impulses of a fallen species—the impulses of the powerful to conquer, enslave, and control those they view as weaker than themselves—underlies every decision that world leaders make. Despite our elites' misguided belief that the world has somehow overcome human nature, building a massive globalized society in which nothing matters but the market, we are getting new evidence every day that this is simply not the case.

For a while it was possible to ignore this evidence and pretend that the neoliberal world order would go on forever. Even in the face of mounting evidence to the contrary, politicians and academics in this country continued to pretend that conflicts between great powers were a thing of the past. Occasionally these politicians would use this arrogant, neoliberal worldview to paint other people as dumb, out of touch, or naive.

Today, few on the political left remember the moment in 2012 when the president of the United States—Barack Obama—openly mocked Mitt Romney during a debate because Romney believed Russia would be a serious geopolitical threat to the United States in the years to come. At one point he famously told Governor Romney that "the 1980s called, and they want their foreign policy back." Obama and his secretary of state, Hillary Clinton, held the absurd view that Russia could be a reliable partner in the future. Clinton infamously gave Russian foreign minister Sergey Lavrov a red "reset" button in 2009, promising the beginning of a new era. This came less than a year after Russia's invasion of Georgia.

Perhaps more infamous were Obama's comments in 2012 to his counterpart that he would have "more flexibility" after winning reelection. Russia—led by Vladimir Putin, who was always lurking in the background—sensed weakness and vulnerability. It swiftly invaded Crimea in 2014 and launched a wide-ranging plan to interfere in our 2016 elections by sowing dissent among Americans. And then, of course, with Obama's vice president in the Oval Office, Russian tanks rolled into the villages of Ukraine under the command of President Vladimir Putin, and it became clear that human nature cannot be suppressed forever, especially when it comes to great-power politics. While the United States has been focused on petty domestic political squabbles, other nations have been planning for the day when American leadership on the world stage ends.

Clearly, Russia under Putin has not—and almost certainly *will* not attained the level of influence or power that the Soviet Union once had. Despite controlling a great deal of fuel in Europe, the country does not have the ability to expand in this way. In that sense, both Obama, who gravely underestimated Russia, and Mitt Romney, who believed it was America's "number one geopolitical foe," were wrong.

The problem is not Russia alone, of course. It is a partnership between Russia and another nation hostile to the interests of the United States—namely, China.

During a meeting with generals from hostile nations, including several from China, Vladimir Putin said that the "unipolar moment" that began at the end of the Cold War would soon come to an end. This belief is also held by leaders of the Chinese Communist Party, who represent a much greater threat to the global balance of power than any nation in history ever has. Just because they are evil doesn't mean they are wrong. And we need to be executing a foreign policy that ensures American security and prosperity, while understanding that our resources are finite, other nations have ambitions, and tradeoffs must be made.

As long as it possesses a massive nuclear arsenal, Russia will be a force that needs to be dealt with carefully on the world stage. China poses a much greater challenge to American interests—and while our foreign policy leadership has been talking about a pivot to Asia for some time, this pivot is rarely actually reflected in policy decisions. The Chinese Communist Party's ambitions are nothing short of supplanting us. While China and Russia could be played off each other during the Cold War, today they share an ambition to upset the global world order. Already, we have seen the devastating effects of this partnership in action.

The Return of History

Vladimir Putin's invasion of Ukraine was a shocking reminder of the cruelty and atrocities mankind is capable of in pursuit of conquest and ambition. In many ways, it can be seen as the opening chapter in the return of history. But it will not be the last, or the most dangerous. Even as Vladimir Putin grows increasingly desperate, killing innocents and destroying cities to further his goal of reunifying the Soviet Union, an even greater challenge awaits us in the Far East.

When Congress began debating how to help the people of Ukraine repel Russian forces, we were constrained in our choices. As we continued to talk about what could be done, our options seemed even more limited. They were limited because of Moscow's nuclear arsenal, which President Putin said he was willing to use. They were limited because nearly all of Europe is heavily dependent on Russia's oil and gas. And they were limited because, as horrible as what Putin has done to Ukraine is and as important as stopping him is, Americans rightly had no interest in seeing their sons and daughters die in the fields of Ukraine. It had taken thirty years, but policymakers were finally being forced to reckon with the limits of American power.

But the limits we face in confronting Russia's expansionist military invasion of Ukraine pale in comparison to the limits we will be under if China continues its ambitions to supplant the United States. In Beijing, we are faced with an adversary that not only has a nuclear arsenal and an influence over global markets that even the Soviet Union lacked, but also has critical control of *our own* key supply chains. Ask yourself this: If China invades Taiwan, would Nike, CitiBank, or McDonald's pull out of China? Or perhaps a better question is whether our government would block imports of Chinese goods as we did goods from Russia. Would our financial elites, so intertwined with China, support the same kind of economic sanctions we imposed on Russia? At this point it is difficult to imagine, which is why any conflict between the United States and China wouldn't just require us to handle the question of nuclear weapons cautiously; it would also require us to withstand the tremendous leverage each country has over the other due to our intertwined economies. It is in nobody's interest—not even China's—for a catastrophic war to break out between the United States and China.

To avoid that outcome, America must be strong, and the Chinese must know it. China can become a rich and powerful country without seeking global domination. Right now their leaders don't appear to believe that; they think America is a declining power that they can push around to achieve their objectives. The path to peace starts with disabusing them of that notion.

For the past thirty years, American policymakers have been blind to the great power conflict between the United States and China that is now before us. We have been primarily focused on a very real threat—the threat of international terrorism—but not the most pressing threat. Going forward, we must be clear about our foreign policy priorities and articulate them to the American people. A great power conflict requires a whole-of-society approach. You cannot impose a whole-of-society response to a threat from above. It requires persuasion, listening, and convincing people that a challenge is worth their commitment.

What is the role of American foreign policy in the twenty-first century? For starters, we must be engaged with the world. There has always been a temptation in American politics to pull back from engagement in the world and to believe we can be safe within our own borders. This is foolish to believe. The Chinese Communist Party is an oppressive regime that commits genocide against its minorities, imprisons its own people, and destroys freedom through the most oppressive and sophisticated surveillance state in the history of the world. If this is how the party behaves domestically in its own country, how do you think it will behave internationally if it becomes the most powerful player on the international scene?

An America engaged in the world, working with our allies, will have to engage the threat of international terrorism. This is a real threat. But it's not the primary threat, and we cannot allow our focus on it to distract from larger priorities. Similarly, an America engaged in the world needs to understand that Russia's leadership has ambitions to reconstitute the Russian empire. Those ambitions, from the world's most nuclear-armed country, cannot be ignored. But while

America will remain engaged in Europe, we are going to need our European allies to step up to the plate and shoulder the bulk of this load. For the fundamental challenge of this century is the rise of China. Meeting that China is going to require the primary attention of American policymakers, business leaders, and citizenry.

Bad Intelligence

The autocratic regimes that challenge us on the world stage have some advantages in competing with us. First among them is their ability to plan long-term and execute on those plans. They don't have to hold town meetings when they want to get something done! Yet that advantage comes at a tremendous cost. First of all, the give-and-take of democracy allows ideas that might have been suppressed to bubble to the top. Vladimir Putin could have used some dissenting opinions in the lead-up to his catastrophic invasion of Ukraine.

Even more importantly, our representative democracy creates the conditions for public trust in our institutions. When you ask people to do something difficult, trust in those institutions is a critical source of national strength. It is something we have enjoyed for much of our nation's history. It existed as recently as twenty years ago when the entire nation rallied around President George W. Bush in the aftermath of 9/11. And it is why I am so concerned that many of the decisions and actions that have been taken in the twenty years since have eroded so much of that trust. The politicization of intelligence. The weaponization of law enforcement. The criminalization of dissent.

We need look no further than the 2020 election to understand just how much damage has been done to public confidence in our vital national security institutions.

Start by looking at what happened when the *New York Post,* our nation's oldest newspaper, broke the news about Hunter Biden's laptop. Almost immediately, America's entire media and political establishment cautioned that this might not be what it appeared. National Public Radio, for example, was quick to describe the story as "discredited by U.S. intelligence." That claim stemmed from a letter by former intelligence officials that warned, without evidence, that "our experience makes us deeply suspicious that the Russian government played a significant role in this case."

In other words: "Trust us."

It was all the national media needed to stop reporting on Hunter Biden's business dealings and what role, if any, his then vice president dad played. It also prompted Twitter, Facebook, and the rest to begin censoring any content related to Hunter's laptop.

Soon later, Mark Zuckerberg would explain how this happened. Months before the election, the FBI had warned social media companies to be on "high alert" about "Russian propaganda" efforts. And of course Russia did work hard to sow dissent and chaos in 2016. However, thanks to the Twitter Files—a deep trove of documents proving the left-wing bias of Twitter—we know there was much less foreign agitation than expected as the 2020 election drew closer. Nonetheless, there was every reason to believe they'd do so again, even if these warnings simply checked some bureaucratic boxes. Russian propaganda is so affective because it plays on our existing fears, divisions, and trends. But just because something exhibits the hallmarks of Russian disinformation doesn't make it disinformation.

One of the men behind that now-infamous laptop letter was Obama's former director of national intelligence, James Clapper.

Clapper, of course, became famous on MSNBC and CNN for asserting that Russia swung the 2016 election for Trump. "Knowing what I know," he said in 2016, "it stretches credulity to conclude that Russian activity didn't swing voter decisions."

In other words, "Trust me."

I've spent the last twelve years consuming intelligence, much of that time while Clapper was supposedly the smart adult in the room. I also spent eighteen months reviewing everything the intelligence community knew about the 2016 election. There is no doubt that

Russia meddled in our 2016 election—they do it in almost every single Western country across the globe. But what is just as clear is that, for partisan purposes, some former senior members of the intelligence community grossly exaggerated what had gone on and fabricated a false "collusion" narrative that dominated America's public discourse for years. To be clear: there has never been any evidence that Russia's meddling in our elections had any impact on the outcome. And there certainly was nothing to indicate that Trump or his team colluded with the Russians to beat Hillary Clinton.

And for that matter, why would Putin have wanted to help Trump win? It was Hillary Clinton who promised to reset relations with Russia. It was her boss, Barack Obama, who mocked Mitt Romney's concerns about Russia and promised to be more "flexible" with Russia after his reelection. And it was the Obama-Biden administration that let Putin waltz into Crimea without consequence and refused to sell the Ukrainians Javelin missiles. Vladimir Putin's ambitions have done just fine when the caretakers of American decline have run our foreign policy.

But we don't need to get inside Putin's head to understand how this all spiraled out of control domestically.

Adam Schiff, the Democrat chairman of the House Intelligence Committee, boldly proclaimed in 2017, "The Russians offered help, the [Trump] campaign accepted help. The Russians gave help and the president made full use of that help." It was a lie, but no one could ask for evidence because Schiff played the "intelligence card," saying he obviously couldn't "go into particulars." In truth, the media didn't need much convincing; they wanted to believe Trump could only have won with outside help. They wanted him to be illegitimate. They wanted him gone. And they were more than willing to help weaponize the intelligence community if it achieved that goal.

The crazy thing is that Schiff knew his comments were little more than a big lie.

Once committee transcripts were released years later, it became clear that the House Permanent Select Committee on Intelligence knew there was no evidence of collusion. In 2017 Clapper told Schiff that he "never saw any direct empirical evidence that the Trump campaign or someone in it was plotting/conspiring with the Russians to meddle with the election." Former Obama attorney general Loretta Lynch later told the committee that she didn't "recall anything being briefed" on anything that resembled collusion.

Nonetheless the media, Democrats, and half of the American people went on believing Trump was only president because he worked with Putin. That did more damage to our nation than anything Putin could have hoped to achieve by trolling us on Facebook and Twitter.

One of our great challenges is to restore that faith, but it won't be easy. People like Clapper, Schiff, and Clinton weaponized these institutions against conservatives. The media was complicit. Democrats in Congress were complicit. Big tech was complicit. And none of them have apologized for getting it wrong. In fact, Schiff and others continue to peddle their big lie with zero consequences.

There will come a point—likely quite soon—when we will need Americans to trust the intelligence the brave men and women on the front lines have gathered. If they don't, we will find ourselves ill-prepared for what comes next.

But trust is earned, not given—especially after two decades of politicized intelligence. We need to hold people accountable for that abuse. That includes preventing them from serving in future administrations and, where necessary, putting them in jail if they've violated the law. We also have to make sure what we do is actually in America's best interest. We don't need our intelligence community monitoring Americans' social media for "disinformation." We need them focused on external threats, especially those coming from China.

We've spent two decades destroying these institutions. It is time we rebuild them the right way before it's too late.

Real Threats

Democrats occasionally talk a good game about being tough on China. In today's political climate, they cannot afford to do otherwise.

But they rarely follow through.

The reasons for this are varied. As always, there is an extent to which the left views any policy that combats China through the lens of American identity politics and political correctness. I won't pretend to know whether they genuinely believe taking action against the Communist regime in China is racist, or whether it is just a pretext to justify defending the bankrupt status quo that benefits their elite donors. But regardless of motivation, the results are the same—a continued weakening of America.

In 2018, for instance, the Trump administration created the China Initiative, which empowered the Department of Justice to counter Beijing's vast espionage campaign against American universities and research centers. In February 2022 the Biden administration shut this program down. They shut it down not because the program was ineffective or because the threat of Chinese espionage had subsided since 2018. If anything, the opposite was true. The Biden administration ended the program because the far-left activists who now control the Democrat Party smeared it as racist and xenophobic. Similar charges, of course, were leveled against President Trump when he took the drastic step of shutting down travel from China in the early days of the Covid-19 crisis. Ironically, in late 2022 President Biden decided to single out travelers from China for enhanced Covid testing protocols, a move China's Communist regime decried as unscientific and racist.

But wokeness—if that is what it actually is—is not the only problem. Usually our elites will ignore the threat posed by China because they and their donors enjoy the post–Cold War economic status quo that led to massive income gains for left-wing communities on the coasts while hollowing out the middle of the country and former industrial cities. For years I have advocated the need for a smart industrial policy to bring back good manufacturing jobs and ensure that we are supporting the levels of research and development in the United States to stay on the cutting edge of important fields like artificial intelligence, quantum computing, and genomics. But when the Democrat-controlled Congress came to power in 2021, they seized on Republican willingness to invest in a critical industry—in this case, semiconductors—and turned it into a gigantic giveaway to our nation's higher education system. Yes, we invested heavily in America's semiconductor industry, but Democrats had no interest in actually combating Chinese espionage. Instead, they created a $100 billion slush fund for research and development, much of which would go to a higher education system that indoctrinates our children with insane ideas and is regularly exploited by sophisticated Chinese espionage campaigns. In essence, a bill intended to combat China's rise will pour tens of billions of dollars into activities that the Chinese are stealing from us. When I introduced an amendment to put safeguards into this bill and use the opportunity to shore up our counterespionage work, my efforts were defeated.

Our response to China isn't just being handicapped by wokeism. It's also being handicapped by outdated economic interests.

When the Trump administration entered into Phase One of its trade deal with China, I said that it would do little more than boost agricultural trade—and even on that, the Chinese have not kept their word—and give Wall Street the green light to further subsidize China's economy. Structural weaknesses were beginning to appear in China's economy, and that was before Xi locked down his entire country for three years because of Covid. Some Chinese banks were struggling with mounting debt fueled by nonperforming loans, many of which went to sluggish but powerful state-owned enterprises focused on long-term growth instead of near-term profit. Trump's trade deal authorized American financial companies to purchase nonperforming Chinese loans for the first time. It would be a desperately needed cash boost for critical Chinese industries. Beijing was happy. Wall Street was happy. And our trade experts were talking about soybeans. Meanwhile, China was able to plow forward in its national effort to displace the United States and dominate 5G technologies, quantum computing, artificial intelligence, advanced pharmaceuticals, and high-value manufacturing. This happened because Beijing deputizes American companies and turns them into China's advocates in Washington. In far too many cases these firms are more interested in appeasing Xi Jinping to maximize their profit margins than doing what is both morally right and good for their country.

Of course, they don't actually consider America to be "their country." The nation that gave birth to them is an afterthought, an embarrassing parent you wished had already passed on, though you can't quite bring yourself to say that out loud.

These are not the patriotic corporations we remember from our past, just greedy ones that always find a way for the ends to justify the means. How else can you explain Coca-Cola's outrage over Georgia's commonsense voting laws, but its willful ignorance of slave labor in its supply chains? Nike championed Colin Kaepernick as the poster child of America's allegedly systemically racist system while it profited from slave labor in China. And then you have Disney, which filmed its new movie *Mulan* in Xinjiang, the very province where the Chinese Communist Party has genocide camps, expressing outrage that Florida schools will not be teaching seven-year-olds about gender identity. In the credits of *Mulan*, Disney even thanked the local government officials who run the genocide camps for their help in producing the film. You simply cannot make it up!

As I said in the previous chapter, there is nothing wrong with companies wanting to make a profit. A company that doesn't make a profit won't be a company for long. That's what companies do in a capitalist system. But by the same token, we have to understand that we will never be able to confront the threat before us if our public policy is built solely on the pursuit of corporate profit, without accounting for what's in the best interest of America.

There was a time when large American corporations not only made a profit but also did so while promoting patriotism, pride in the values enshrined in the Declaration of Independence, and respect for the dignity of every person. But today we live in a world where many of our most successful companies—which have addresses in America but consider themselves "citizens of the world"—simultaneously defend or ignore Uyghur slave labor, censor conservative voices in the United States, and hold compulsory struggle sessions at their offices in service of the woke agenda.

We are not going to be able to address the unprecedented threat posed by the Chinese Communist Party as long as this White House and American politicians continue to prioritize the whims, pet causes, and speech codes of progressive identity politics above America's economic and national security.

Because China is no longer hiding its strength and biding its time.

Since 2012, Xi Jinping's words and actions have made it clear that Beijing believes it now has enough power to begin remaking the international order in its image, and that the time has come for China to reinstate itself as the Middle Kingdom, the dominant power in the Indo-Pacific and, eventually, the world.

China now pursues economic imperialism, entrapping the developing world through the exploitative loans of its Belt and Road Initiative. And it is now an imminent aggressor to our allies and partners in Taiwan, Japan, India, and elsewhere. These trends won't get better—they will only worsen and accelerate from this point forward. There are several actions we must take immediately to address them.

First, it starts with unity and clarity about the threat we face. The gravest threat facing America today, the challenge that will define this century and every generation represented here, is not climate change, the pandemic, or the left's version of social justice. The threat that will define this century is China. And we will need a whole-of-society—not just government—effort to fight this threat.

Conservatives need to understand this. Liberals need to understand this. Small businesses need to understand this—and so do businesses like Tesla and Amazon. If these megacorporations won't get on board, we need to start asking ourselves why they deserve the protection and patronage of the US government, if they continually promote and defend efforts that undermine our national security and long-term economic viability.

Second, we need to empower our government to counter Beijing's infiltration. Those are not words you often hear coming from someone with an "R" beside their name. But being a conservative is not being antigovernment. It's understanding that most of the answers to the problems in life don't come from the government, but there are a few things that the government has to do—and one of them is to provide for national security.

And in that vein, we must begin decoupling key industries from China. We cannot rely on Beijing for rare earth minerals or pharmaceuticals. We cannot continue to collaborate with them on sensitive, groundbreaking research. We cannot allow American retirees to unknowingly fund China's military. And we cannot allow Chinese technology to operate inside our country.

Third, we need to revitalize our industrial capacity if we're going to be able to make this an American century, rather than surrender it to Communists. A nation dependent on hostile regimes is not going to last long. You can't be a great power if you're not an industrial power. You have to be able to make things. That's why in this country we made a decision a long time ago to buy our weapons, particularly our airplanes, from American companies that make them in America.

Well, the menu of things critical to our national security has expanded. I would argue that semiconductors and the active ingredients in our pharmaceuticals are just as important to our national security as weapons. Relying on a hostile adversary for these things and more will leave us vulnerable and weak.

Think about the hand-wringing we saw over banning Russian oil, which accounts for a small percentage of America's consumption and can easily be replaced, four times over, through increased domestic production. Now imagine, some years from now, that we're in a conflict with China. Think about what would happen if they declared, "We're just going to cut you off of everything—lifesaving drugs, iPhones, the rare earth minerals you need to power your weapons systems."

Think about all the things that we depend on China and its manufacturing capacity to provide. Imagine being cut off from that. Imagine the leverage that would give them. You think our options are limited now with Russia—we wouldn't have many in that conflict. It should be obvious to everyone by now that our economic dependence on Beijing is a vulnerability we can no longer accept.

Finally, we need to empower our allies and partners. This is not just a competition between China and America. Beijing seeks dominion over its neighbors. It views them as vassal states, tributary states. That's its vision for the future of the Indo-Pacific region. These aren't buffer states—there are no buffer states—these countries just happen to be on the front lines. In the coming months and years, our alliances and partnerships with Taiwan, Japan, Australia, Korea, India, Brazil, and others will be more crucial than ever.

If our European allies are to stand firm against Beijing as well, they will need to be more skeptical of China's economic overtures. And most importantly, they will need to take greater ownership of their security—so they can take a leading role to counter Putin's aggression, and so we can focus on the threat of Communist China in the Indo-Pacific.

The horrific invasion of Ukraine has made—or should make—countries across Europe realize they are not living in some sort of postconflict utopia. But that realization cannot be a momentary blip that evaporates once Putin loses. It must be sustained so America can prioritize, and direct its resources to effectively counter Beijing in the years to come.

The Confidence to Prevail

In the course of human history, the American experience is the exception, not the rule. Almost everyone who has ever lived did so in cultures, societies, or nations where their rights were what those in power allowed them to have. And where a tiny minority always stayed on top, and no one else even had a chance.

Three times in the last century America attempted to create a world order that reflected our exceptional experience.

First, after World War I, President Wilson's dream of a League of Nations collapsed, leading Americans into an era of isolationism that allowed the rise of Nazis in Germany, Fascists in Italy, and militarism in Japan to plunge the world into a bloody second world war.

Then, after the end of that war, an Iron Curtain ushered in almost five decades of cold war that left the world divided between prosperous and free democracies on one side and, on the other, stagnant and bankrupt Soviet satellites that eventually collapsed.

And finally, after the end of the Cold War, America and other democracies believed we had reached the "end of history." They opened up our markets, resources, and technology to China, and allowed it to cheat, lie, and steal, because they believed that once China became rich, it would become more like us, free, prosperous, and a responsible member of a liberal world order.

And now we are reaping the decades of decadence that we sowed.

Even though the old, failed bipartisan consensus on China has collapsed, the status quo policies of it remain largely in place. The collapse of an old strategy alone will not produce a new one that works.

This is not just a foreign policy issue. It is tied to virtually every single major domestic issue we are facing.

Producing a new strategy that unites and mobilizes not just Americans but the democracies of the world to confront the threat of China is THE central issue of our time. If we fail—if strategic paralysis leads America to isolationism and leaves democracies fractured—then tyrants in Beijing, Moscow, and Tehran will become more powerful and aggressive. We will enter into a new dark age.

And ultimately, we will confront the most horrific open conflict in the history of man.

If we want to stop this, we must begin the hard work of rebuilding trust in our institutions. For we will have challenging times ahead. Most importantly, I fear China's desire to dominate Asia and the impact that would have on the world. Today, Asia contains approximately 50 percent of world GDP. A world in which China dominates Asia is one in which it controls the rules of the road of global commerce. And if you look at how it treats its own citizens at home, or the people of Hong Kong, you can imagine how the Chinese Communist Party would treat the rest of the world if it had the power to call all the shots.

Beijing is serious about dominating Asia. It's already made this clear through its "salami-slicing" strategy in asserting claims over the Senkaku and Paracel Islands, among other places. But if you're looking for a sign that China has decided it's ready to go all out in establishing control over the entire continent, the first place to look is the island of Taiwan. China desperately wants to incorporate the island into its version of One China. For over forty years American policy has been that the question of Taiwan's status needs to be peacefully settled between China and Taiwan. In the meantime, we provide arms to help Taiwan defend itself against Beijing. One reason America's support of Ukraine has been important is that the leaders of the Chinese Communist Party have been watching very closely what happens there as Putin seeks to absorb Ukraine by force.

Taiwan is a hugely important global issue for a number of reasons. First, the whole world knows that if China were to overtake Taiwan by force, then America would have allowed a strategically vital ally to whom we have made assurances to be swallowed up by Communist China. What message would this send about our reliability as an ally to important countries like Japan, South Korea, Australia, India, and Vietnam? All of these nations sit at the doorstep of China's ambitions. All of them in their own ways are showing courage in resisting the expansionist desires of China. China has border disputes with most of these nations as well.

Further, Taiwan is strategically important because of its location in the famous sea lanes of the Pacific Ocean, through which nearly half the world's container ships travel. If Beijing dominates these, it will have a chokehold over critical goods and the spread of international culture. In a Chinese-dominated world, if the United States does something the Chinese Communist Party doesn't like, Beijing will be able to simply block sea lanes to US ships and starve us of medicines, metals, and minerals. And if you think Communist China wields too much control over American culture as is—regularly coercing Hollywood writers to erase characters and plotlines it doesn't like from US films—then imagine a world where Beijing is fully in control. But even beyond hypotheticals, Taiwan is a tremendously important part of the global supply chain and a huge producer of semiconductors, one of the most strategically important items of commerce today. Deterring China from starting a war in Taiwan is tremendously important if America is to live in peace and prosperity. But at the moment, most Americans don't see it that way; either they don't know, or they simply don't care.

The only way we can credibly deter China is if American leaders have the confidence of our own citizenry. We need to be able to clearly explain why America's commitments to Taiwan are important to uphold. The American people need to have confidence that our foreign policy leaders are making prudent judgments based on the realities of the world that we live in. Among these realities is that America cannot be the world's policeman. Our power and influence is not infinite, and not everything that goes wrong in the world is a matter of America's most pressing national interest.

That confidence is necessary because if America is to deter China from invading Taiwan, it must continue to support the Taiwanese with smart arms sales. The confidence of the American people is critical in ensuring that Congress approves those arms sales.

The confidence is necessary because credible deterrence will require the Chinese to believe that Taiwan has the support not only of America but also of critical allies in their region. And those allies need to know that the American people back the commitments our nation is making.

And the confidence of the American people is necessary because deterrence only works if China knows that America is willing to go through with difficult things if they do invade Taiwan. I hope those difficulties don't include a war—because war would be catastrophic. But they will certainly include economic pain and a world that looks radically different the morning after China invades. America will only have the strength to lead in that world if the American people feel informed by their leaders and, most importantly, confident in the judgments they are making.

I understand why so many Americans have their confidence shaken. Hubris and a misreading of the world as it is have done tremendous damage to our nation's institutions. But I believe it is possible—no, existentially important—to rebuild them.

Chapter 5

THE RISE OF THE EXPERTS

"The Most Successful Fiscal Policy in the History of the United States Government"

Today, most people can probably recall the exact moment when they realized Covid-19 was going to change life in the United States for a very long time. For some, it was when they picked up their iPhones in March to find that the NBA had canceled the remainder of its season; for others, it was when a news alert came through, announcing that Tom Hanks had come down with the virus.

I will always remember the day it all came together for me. It wasn't only that a novel coronavirus was spreading through China; I had known that for months. There was good public reporting on this everywhere—the first public source I saw was Bill Bishop's excellent China newsletter. At one point in March, I remember texting a group that included my wife and our closest family friends with an ominous message.

"The world," I wrote, "is never going to be the same."

But before I could worry about the world in a larger sense, I was worried about the United States, primarily my constituents in Florida.

When Dr. Fauci and the White House Coronavirus Task Force announced a fifteen-day shutdown of the country, intended to stop our health-care system from becoming overwhelmed by too many cases of Covid-19, I had just flown back up to Washington, DC, to vote for what had become known as the second relief package.

Whatever came next, I knew it was going to be bad.

My first thoughts went back home to South Florida. In my hometown alone, I knew hundreds of small business owners who wouldn't be able to survive fifteen days without revenue. I thought of my constituents who ran restaurants and clothing stores, for instance, which operated on very thin margins and often teetered on the edge of insolvency even without a global public health catastrophe. Small businesses were going to bear the brunt of these lockdowns. We sat on the cusp of an economic crisis of historic proportions. I knew I had to do something, and luckily, I had a pretty good idea of what it was.

During that session of Congress, I was the chairman of the Senate Small Business Committee. For months, my staff and I had been creating a program designed to help business owners who had run into issues with supply chain interruptions in China. Officially, the program was a reform of the Small Business Administration, or SBA, which at the time was the smallest cabinet-level agency in the federal government. I wanted to bring the SBA back to its roots as an agency that could help banks, especially small community banks and other lenders, serve the small businesses that formed the bedrock of their communities by supplying essential goods and services and employing workers in good and stable, long-term jobs.

Under the plan we had drawn up over a year earlier, the Small Business Administration would have played a major role in seeding manufacturing companies with low-cost capital to expand their manufacturing, supply chains, and domestic workforces in America. Covid hit right in the middle of this process, and in doing so exposed a major problem with the way our elites had come to think about the economy. Namely, that our government acts as if we have unlimited resources—not only hard resources such as money, but also more abstract concepts such as trust in our institutions and national unity. Over the past few decades, our government seems to have adopted the notion that we can penalize, politicize, and exploit businesses, communities, institutions, and citizens whenever it seems necessary in the short term, and the long-term consequences will be negligible.

This is what happened when we decided to close schools for an undetermined period of time during the pandemic. It's what happened when we brought China into the World Trade Organization. Because we had enjoyed prosperity and stability for so long, we assumed it would go on forever—that the world would look the same once we decided to start paying attention to it again.

But the Covid-19 pandemic was a rude awakening to the experts who assumed that spreading our supply chains all over the world would have negligible consequences for American communities.

Right away, the virus brought the kind of supply chain disruptions that my team had warned about when we began crafting our SBA expansion. With China locked down, thousands of products were now effectively unavailable. This included not only masks and key medicines—a shocking number of which were, and still are, produced in China—but also the kind of things we bought and sold every day: clothes, semiconductor materials, car parts, and more. This revealed the hidden costs of spreading our supply chains out across the world—costs that seemed worth it during the three decades of hyperfinancialization that preceded Covid.

But as the lockdowns worsened, it became clear that this crisis was something deeper than a supply chain crisis. It was a matter of survival. Small businesses were being forcibly shut down. It became a macroeconomic and, potentially, global financial crisis. The bottom began to drop out of the stock markets, with no prospect of real economic activity to pick up the slack. Usually recessions take months to take hold. This one took days. Friends I knew who ran businesses—first in hotels and then nearly the entire commercial real estate market—told me they faced defaults. No one had faced a situation like this before. We had been thinking big with my SBA supply chain plan. We needed to think bigger.

I moved my laptop two floors up the Russell Senate Building to the Small Business Committee offices. Most senators usually have jam-packed schedules in fifteen-minute blocks, filled with meetings with constituents, other senators, staff, and otherwise. With everything shut down, none of that was happening. I opened up the doors to the Small Business Committee room, and we basically held court. Over the course of less than two weeks, through around-the-clock meetings with small groups of senators from both parties, our staffs, and Trump Administration officials fueled by Chipotle and cafecito, we had a new plan.

The plan was to build from the foundation of the SBA's existing network of small-business lenders to create an emergency lifeline for small businesses. But instead of loans, the lenders would be giving out government-backed grants tied to keeping workers on payroll, keeping their lights on, and keeping their businesses operating despite Covid restrictions. I called it the Paycheck Protection Program, or PPP.

PPP would go on to provide over $800 billion to the United States economy in support of small and midsize businesses that kept their workers on payroll during Fauci's lockdowns—the largest amount ever spent through a single federal program in such a short period of time. But it almost didn't happen.

Just as the public health experts were wrong about Covid, the economic experts were wrong about the economy. In these meetings and elsewhere, I was told by economic experts—including some of the same people who were involved in the 2008 bailouts—that PPP was economically "inefficient." You see, the most efficient strategy, according to these experts, for businesses during the pandemic would have been to lay off workers immediately. If businesses didn't have any work for their workers to do, then it would be inefficient for the government to help keep them on payroll. Plus, these experts said, the Covid economy was a new economy. People would buy more online from Amazon instead of going back to their local small businesses. The lockdowns were just accelerating an inevitable process. Besides, they said, for public health reasons it was better for people to stay at home rather than to work.

I can tell you with great confidence what would have happened to the Covid economic response in March 2020 without PPP. We would have seen the 2008 bailouts again, in a new form. The Federal Reserve would have rained subsidized credit on big companies who could afford to weather lockdowns—or, as we saw with companies like Amazon, even benefited from them. Workers would have all been put on unemployment. And small businesses would have been told to take out more debt or just file bankruptcy on their way to the unemployment line. The cascading effect on families and communities would have been devastating. And it would have rippled through the property markets and ultimately the financial markets, just like it did in 2008.

I told other senators and members of Congress, many of whom are small business owners themselves, that it would be a disaster. They got it. Workers put on unemployment would become detached from the workforce for much longer than the pandemic. Small businesses would close their doors permanently. We would see the most significant concentration of wealth in the hands of the few, the experts, in the history of our country. That was unacceptable.

In the end, the Paycheck Protection Program proved to be the single most effective relief program ever passed by Congress. We know this not just from the overwhelming anecdotal evidence from small business owners but from actual data. The Senate Small Business Committee reports that "the macroeconomic trends surrounding [PPP] are clear: employment was falling, and then, after PPP started, employment started growing again." Former Congressional Budget Office director Doug Holtz-Eakin went even further, calling PPP "the single most effective fiscal policy ever undertaken by the United States Government."

Since we passed the program, I've heard firsthand from small business owners across the nation who said that PPP was a life raft that helped them navigate the pandemic's tumultuous waters.

In May, economists expected the economy to lose upward of 8.3 million jobs and unemployment to reach levels unseen since the Great Depression. Instead, with more than $500 billion in PPP loans injected into small businesses across the nation during April and May, our economy added a record 2.5 million jobs—all amid a pandemic.

Across the United States, the PPP helped support up to 55 million jobs, including up to 4.5 million in manufacturing, with an average firm size of just twenty employees. Even before the pandemic, manufacturing was experiencing a widening skills gap. An inability to retain skilled employees through the downturn would have exacerbated that gap and further degraded American global competitiveness.

The PPP acted as an emergency brake on further decline. It meant overriding the "free market," leveraging local banks and other financing institutions that were deeply integrated with their communities, and injecting billions in federal investment toward a clear common good: keeping American workers attached to their workplaces. Despite cynical attacks—such as President Biden absurdly comparing the program to his student loan bailout for educated elites—the program's benefits swept across businesses in every industry and corner of our nation. And while many in the media have tried to focus on fraud, the benefits of the program are undeniable. But then again, it should come as no surprise that the national media is openly hostile to a program that saved millions of mom-and-pop businesses. Those small business owners don't have the funds to advertise in the *New York Times*, and they don't have the time to read the latest drivel from the *Washington Post*.

Looking back on the experience of writing and implementing PPP, a few things are inarguable. The first is that the program staved off catastrophic economic consequences for this country. The second, and perhaps the most significant as we look toward the future, is that the program—which, again, will rank among the most successful ever implemented by the US government—was written against the advice of virtually every expert to whom our leaders would typically turn to make these decisions. If we had gone to the unelected men and women who crunch numbers and effectively make policy during stable, prosperous times, the results would have been disastrous.

Unfortunately, the US government has come to rely more and more over the past few decades on these unelected experts, often implementing policy based on their advice that is completely wrong-headed.

The Technocrats

In the United States, trust in our institutions has always been a source of national strength—and an advantage of representative government over autocracy.

From the time our children are young, they are taught about the system of checks and balances that holds our nation together. Most of these children can tell you about how Congress checks the president, the president checks Congress, and both check the judiciary, striking a balance that makes this country unlike any other in the world. At some point, they'll usually tell you that at the bottom of this system, you have democracy—people coming together to vote for representatives, who in turn write and enact legislation that benefits the people who voted for them.

But over time, we have changed from a nation governed by "We the People" to one governed by experts. Broadly speaking, this trend began in the 1930s, when President Franklin Delano Roosevelt—after threatening to pack the Supreme Court with friendly justices if he didn't get his way—managed to ram the New Deal through Congress. Almost overnight, the United States saw an explosion of new federal agencies, each of which had an independent, supposedly apolitical expert at its head.

Throughout the twentieth century, Congress and the executive branch came to rely on these supposedly independent bureaucrats more and more. As the government became involved in more aspects of our daily lives, the need for more government "expertise" exploded. And as the federal bureaucracy grew, so did the number of unelected experts who had the power to influence policy. Supposedly, these people were guided by facts and data alone—in time, they came to be known colloquially as "technocrats"—and it was generally assumed that politics did not factor into their decisions.

For a while, most people probably didn't notice their new technocratic overlords, especially when it came to things like highway projects and health-care reimbursement rates, both of which have recently come to be governed not by the will of the people but by small groups of experts who are supposed to be guided only by data and objective facts. At the Federal Reserve, which is supposed to be an apolitical institution, groups of technocrats who are supposedly guided by nothing but data and cold hard facts make political decisions every day, often at the behest of the president and the financial sector. Rather than humbly keeping prices stable in times of economic turmoil, the Federal Reserve has created an increasingly exotic suite of financial programs that function as subsidies to Wall Street. For nearly a decade after the 2008 financial crisis, the Federal Reserve engaged in unprecedented purchases of boutique Wall Street assets that broke from its long-standing practice of being a neutral market player. During Covid, the Federal Reserve took congressionally appropriated taxpayer money it was supposed to use to save small and medium-sized businesses, and instead used it to buy the bonds of large corporations. The Federal Reserve spent valuable resources it could have used to gather more information about the costs of living everyday families face, but instead it launched a series of events to discover "racism" in the economy.

As we saw during the early stages of Covid, there is great danger in allowing unelected bureaucrats to begin dictating policy. It is the job of senators, members of Congress, and presidents to dictate policy—that's what their constituents elected them to do. If we make bad decisions, people can vote us out of office. But the thousands of people currently on staff at agencies such as the Department of Education or the Environmental Protection Agency were not elected to anything. Many of them weren't even *appointed* by anyone who was elected to anything. In most cases, they are employees who are supposed to perform a very specific and nonpolitical function for their agency. But now, you have the Department of Homeland Security—which is supposed to defend our nation—working with Big Tech companies to censor what it considers misinformation on everything from the origins of Covid to what happened with Biden's disastrous withdrawal from Afghanistan.

In recent years, we have seen what happens when these agencies become politicized—and when experts begin pushing their personal beliefs on the American public under the guise of "science" or other "expert guidance." There is no better example than the various public health officials in our federal government, many of whom rose to prominence during the Covid-19 pandemic. But there are many others whose names we don't know. In many cases, these are people who staff the ranks of organizations such as the Department of Health and Human Services and the National Institutes of Health, both of which regularly issue guidance to American citizens about diseases, health care, and potential treatments.

Often, this guidance is backed by good science and sound reasoning. There is no reason to assume that these entities have become completely politicized, or that we shouldn't trust anything they say. But if the last few years have shown anything, it's that we should be extremely skeptical when these organizations issue guidance about hot-button political issues.

Our country's Covid response was doomed to failure because we were too slow in transitioning from an initial crisis response mindset to one that was a more sustainable balance of the crisis with the need to live our lives. But the situation was made far worse because our public health officials took it upon themselves not only to advise elected policymakers but also to use their public platforms to try and make policy themselves. They often lied to the public to get people to do what they wanted. During a television interview that would soon become famous—though not famous enough, in my opinion—Dr. Fauci explained why most people did not need to bother wearing masks. "There's no reason to be walking around with a mask," he said. "When you're in the middle of an outbreak, wearing a mask might make people feel a little bit better and it might even block a droplet, but it's not providing the perfect protection that people think that it is. And, often, there are unintended consequences—people keep fiddling with the mask and they keep touching their face."

Later, he would say that he was discouraging the use of masks not for any of the reasons he mentioned on television, but to make sure that first responders and frontline workers would have enough masks to do their work. He might have been right that making sure that first responders and frontline workers would have masks was a more important public good than the right of the American people to make informed decisions about their health. But it wasn't his decision to make. And lying to the American people about your motivations, especially as a government official during a once-in-a-generation pandemic, *is* the fastest way a public official can destroy trust. Once government officials begin telling the American people lies about their work—even if those lies are small, and even if they are meant to protect people or to bring about some social good—then those government officials, and by extension *all* government officials, lose credibility. As a government official myself, I would very much prefer that this does not happen. It makes it much harder to persuade Americans when there is a need for national unity and sacrifice.

Of course, just like too many American corporations, Fauci and the rest were totally focused on the short term and never once considered the long-term impacts their decisions would have on our nation. But that is not sustainable. As we've seen with corporate America weakening our nation, lies and misdirection from public health officials cause people to lose trust not only in specific people but also in the institutions that those people represent. This is another reason that having one publicity-loving official declare that he is the sole representative of an entire sector—"The science," in this case—is so dangerous. After enough flip-flopping on mask guidance, lockdown rules, and other matters of "settled science," the American people will begin (quite rightly) to distrust anyone else who claims to speak for science.

In the early days of the crisis, we abandoned the very ideals that separate our government from organizations like the Chinese Communist Party. We decided—or, rather, we *didn't* decide; these unelected experts decided—to forgo open debate, democracy, and representative government in favor of edicts issued by a small group of unelected government bureaucrats, and all of us are still paying the price for that decision today.

Sometimes our experts and mainstream media went so far as to *praise* the Chinese Communist Party for the programs they put in place to handle the pandemic. In April 2020 the *Nation* declared that it was "chastening to note that whereas China under Xi has suppressed the latest coronavirus at the human cost of three lives per million population, the United States under Trump is still struggling to overpower it, having already sacrificed 145 of every million Americans."[9] Shortly thereafter, Reuters reported that "China's total reported death toll is below 5,000 and new infections are negligible, the result of draconian lockdowns, millions of tests, and strict contact tracing that set the stage for an economic rebound."

It's no wonder that the experts in the United States—many in the same circles as the people who had been drawing us close to China for decades—adopted policies that so closely emulated those of President Xi's.

What's more amazing, of course, is how many people went along with the decisions that these unelected bureaucrats were making. Few bothered to ask where they were getting their data, how they were arriving at their conclusions, or—most importantly—what the consequences of their orders would be in one, two, or ten years. I cannot count the number of times I was told to "trust the science" and be more considerate of other people. Of course, it wasn't long before liberal elites began using "Trust the science" as a kind of mantra—one that quickly came to mean "Shut up and do as you're told." When normal citizens questioned the wisdom of wearing masks outside, which seemed ridiculous given the "science" on how the virus spread, they were told, in effect, to shut up and do as they were told. Then, a few months into the pandemic, when the "science" changed and the government issued new guidance that acknowledged the uselessness of outdoor masking, no one issued those people an apology. The government did not adopt a new tone of humility when issuing more guidance, all of which it claimed to be backed by "the science." If anything, they grew more strident in their belief that they were right and the people they ruled were wrong.

In part, these experts were slow to adapt their guidance because the rules they were issuing almost never applied to them. The rules certainly didn't apply to the local officials who were in charge of implementing the guidance of public health agencies at the level of the city and state. By now, it's almost a cliché to list all the liberal mayors and governors who were caught flouting their own lockdown rules while ordinary people were forced to remain in their homes. I'm sure everyone remembers Governor Gavin Newsom of California, who in 2020, while the rest of his state wasn't allowed to dine indoors, made an exception for himself to attend a birthday party for a lobbyist at the luxurious French Laundry, the fanciest restaurant in America. Or when House Speaker Nancy Pelosi got her hair done at a San Francisco salon despite hair salons in that city having been forced into closure for almost six months.

By almost any measure, the lockdowns were a failure. Looking at the most recent data, which accounts for age and comorbidities, the death rate in my home state of Florida—which lifted restrictions earlier than any state in the nation and refused to clamp down further even when every expert in the government was advising otherwise—is almost identical to that of California, the state that had the most restrictions and enforced them most stridently. The efficacy of the vaccine—which we were assured repeatedly by the experts would stop infection and transmission—is also in doubt. Never in our nation's history have we had such a divide in trust between those in positions of authority and the governed. In the roughly two years since the White House Coronavirus Task Force announced its "15 Days to Slow the Spread" plan, the American people have come to distrust unelected experts more than ever. When Dr. Fauci announced he would be stepping down in August 2022, few Americans were sad to see him go.

Unfortunately, Fauci is just one of the many unelected experts who have come to dominate American life. We only know his name because the left, in their usual way, made him into a kind of online meme and praised him in their publications whenever they got the chance. The other unelected experts have the potential to do far more damage.

In some cases, they already have.

Irreversible Damage

In general, there is nothing wrong with experts, especially when it comes to complex fields such as hard sciences and medicine. When making decisions that involve these areas of expertise, elected officials must consult people who have spent years studying these things.

The problem occurs when those experts form groups, and when those groups begin to banish all dissent. This becomes especially problematic when most members of these groups begin to lean in one political direction.

Sadly, this is exactly what has happened in the United States, particularly when it comes to the extremely important field of pediatric medicine. In recent years, we have seen several stories every day about how to deal with the extremely contentious issue of children who claim to be transgender. This issue, which seems to have been spurred on at least partly by the explosion of media coverage about it, has nonetheless come to affect many American families.

In a perfect world, we would decide how to handle it through careful debate and consideration of the evidence on both sides.

That is not what has happened. Instead, we have begun looking to organizations such as the American Academy of Pediatrics, or AAP. According to this group, more than two-thirds of whom are registered Democrats, children can begin transitioning "at any age."[10] According to an article in the *Free Press*, the AAP has begun "deferring to small, like-minded teams of specialists ensconced in children's hospitals, research centers, and public health bureaucracies, rather than seeking the insights of pediatricians who see a wide cross-section of America's children."[11]

The results have been troubling, to say the least.

In recent years, the AAP and other similar organizations have expressed support for "gender-affirming care" for young people. When today's left-leaning public health experts say "gender-affirming care," of course, what they're talking about is encouraging children—some of whom are as young as five years old—to undergo sex change procedures, often through the ingestion of untested drugs and, eventually, surgeries that will leave their bodies permanently altered.

One of the first steps on this journey is puberty blockers. As the name suggests, these are drugs that are given to children to stop the onset of puberty. They are federally approved for treating adolescent cancers and other severe diseases. As of this writing, they are *not* federally approved for delaying puberty in children experiencing gender dysphoria. Yet the Biden administration is actively promoting that off-label use as part of a "gender-affirming care" regimen. Unsurprisingly, the same experts who told people not to get within a mile of drugs such as ivermectin because they were untested often have absolutely no problem with giving children untested drugs to stop their natural development.

In some cases, they actively encourage it. According to a document published in March 2022, the public health experts in the Biden administration claimed that "gender-affirming care improves the mental health and overall well-being of gender diverse children and adolescents."[12]

This is shocking for many reasons, primarily that the authors of this document seem to have accepted the premise that children who experience gender dysphoria *should* take action and halt their development. In their eyes, this isn't even a question. But even if we do accept this assumption—which we should not—it does not follow that the best way to bring about this outcome is via untested drugs.

If nothing else, shouldn't we at least make sure the drugs are safe?

This is the question I asked Dr. Lawrence Tabak, then the acting director of the National Institutes of Health (NIH), in May 2022. I asked him where the safety data was on puberty blockers in children experiencing gender dysphoria. I asked where the data on effectiveness was, as well as the data on long-term risk. While asking these questions in the Senate, I recalled that when it came to ivermectin, I didn't need to ask any of them. Then, the public health bureaucrats like Dr. Fauci were happy to do it for me. But when the politics shifted, the questions suddenly stopped.

In response to my questions, Dr. Tabak said, "NIH funds a small number of observational studies." He said that "over time, [the NIH] will be able to better answer the types of questions you're posing."

In other words, there are no answers. We have absolutely no meaningful data regarding the long-term implications of giving these radical, life-altering drugs to elementary school children. We cannot possibly weigh the costs and benefits. The Food and Drug Administration is nowhere close to approving such uses. Yet the Biden administration, corporate media, and so-called experts are actively promoting them. It's irresponsible and dangerous.

For example, what happens to an eleven-year-old girl who starts taking puberty blockers and then decides at age sixteen that she wants to live as a woman? The government technocrats of the Biden administration—speaking, as always, as if they have all the data in the world backing them up—have said that the effects of puberty blockers are reversible. In the document released in March, the second and final page includes a chart that explains what different treatments for gender dysphoria are, when they are used, and whether they are reversible. Although other parts of the document contain footnotes and links to different medical studies, this section includes neither. The authors simply include the word *reversible* beside puberty blockers and expect that the American people—who they believe have been conditioned to accept anything that appears in a government document of this sort—will believe them.

But based on Tabak's sworn testimony, it is clear that this particular claim is misleading at best. A major Swedish institution admitted in May 2021 that the off-label use of puberty blockers in adolescents could lead to "extensive and irreversible adverse consequences such as cardiovascular disease, osteoporosis, infertility, increased cancer risk, and thrombosis."[13] This came on the heels of a major study from the United Kingdom, which found that there was "'very low' evidence of benefit to allow children with gender dysphoria to have their natural puberty blocked." In response, the Swedish institution declared that moving forward, no children under the age of sixteen would be given puberty blockers. England took a similar step last year, noting the "scarce and inconclusive evidence to support clinical decision making."

In the United States, where ideology too often guides science, we have made no such progress. Instead, the Biden administration has doubled down on its support for "gender-affirming care." Again, the clinical, sanitized language of this term suggests that it is a routine, well-studied set of practices that have been around for decades. Nothing could be further from the truth. When doctors talk about gender-affirming care, they are talking, in many cases, about irreversible procedures that often pose great danger to the people who undergo them.

When a woman undergoes surgery to become a "man," for instance, the surgeon will often cut tissue from the patient's forearm or thigh to create a new body part; sometimes they will cut into the patient's vaginal tissue and "make it longer, turning it into a defined phallus."[14] Surgeons will also cut into the patient's breasts and remove enough tissue to flatten the chest, then sew up those incisions, leaving two large scars below the pectoral muscles.

When you describe these procedures as what they are, the details seem gruesome. It is no wonder that the bureaucrats in the Biden administration continue to invent new, safe-sounding words for the procedures they are pushing on children.

Some of these procedures have been performed for decades, but only on patients who were fully grown and aware of the risks. They were also performed with much less frequency. One surgeon quoted in the *Washington Post* said that by 2018 he was doing "about 300 procedures a year, whereas it was only about 50 in 2000."[15] Several studies in the years since confirm that the number of transgender surgeries in the United States is on the rise.

And it has not shown any signs of slowing down. In 2021, a medical market research company published a study on the size and potential growth rate of the "U.S. sex reassignment surgery market." According to their research—which was only about the numbers underlying the enterprise, and thus apolitical—the market is "expected to expand at a compound annual growth rate of 11.23 percent from 2022 to 2030." The study also quoted an article from the *New York Times,* which found that "around 1.4 & 1.3 percent of young people aged 13 to 17 and 18 to 24, respectively, are transgender in the U.S."[16] But that number is growing. According to a recent report, the number of young people in the United States who identified as transgender doubled between 2017 and 2022.[17]

Clearly, this is an issue that demands our attention. It certainly demands that we hold the people pushing these procedures on our children—who, according to several studies, claim to be experiencing "gender dysphoria" at increasing rates—to account for their politically motivated recommendations. We know that the majority of individuals with gender dysphoria, especially young ones, eventually lose their desire to identify with another sex. This has been clinically documented in several places, including a study published in the International Review of Psychiatry in 2016. But Democrats, including the woke public health officials who make up our scientific establishment, now claim that this study is "flawed." Of course, these same activists cannot point to a study that *defends* poisoning our children with off-label drug use, or mutilating them with dangerous surgeries.

It doesn't take long to find examples of people who deeply regret undergoing what is now known as "gender-affirming care," especially among people who underwent this care when they were young. In June 2022 the *New York Post* published a long article that included many stories from children who, in their own words, had been "failed by the system." The stories of these people are extremely troubling to read, especially when considered in light of the fact that the Biden administration continues to push "gender-affirming care" on young people. In almost every case, these young people seem to have been swept up in the excitement about gender dysphoria on social media sites such as Tumblr and Instagram.

One of these young people, who transitioned from a girl to a boy at the age of fourteen, recalls that she was "going through a period where I was really just isolated at school, so I turned to the internet. My dysphoria was definitely triggered by this online community. I never thought about my gender or had a problem with being a girl before going on Tumblr . . . The community was very social justice-y. There was a lot of negativity around being a cis, heterosexual, white girl, and I took those message really, really personally."[18]

Reading this testimony, it is hard not to think of the Biden administration's stance toward social justice issues—one that views "whiteness" as a historical evil that needs to be rooted out of society, and the United States in general as an oppressive society that was engineered by white supremacists to oppress racial and sexual minorities. It is no wonder that there is very little difference between the bureaucrats who staff this administration and a bunch of confused children writing in internet chat rooms. They have the same views on almost everything—including "gender-affirming care."

As with Covid, it appears that for woke ideologues, "following the science" means getting the preferred outcome regardless of the evidence. Anyone who presents contrary evidence, in their view, needs to be silenced. When the writer Abigail Shrier, who interviewed dozens of people who had come to regret their childhood transitions, wrote a book about the phenomenon (aptly titled *Irreversible Damage*), dozens of public health officials spoke out against her. In time, the outcry grew to such heights that the retailer Target pulled the book from its shelves and apologized for ever carrying it in the first place.

We need to have a higher standard when administering life-altering drugs to children. As usual, Florida is leading the way in establishing that standard. Florida surgeon general Joseph Ladapo cited "the lack of conclusive evidence" and "the potential for long-term, irreversible effects" in his memo blocking hormonal and surgical interventions for children with gender dysphoria.

The left railed against that memo, but what Florida is doing is common sense. What is crazy is that the same people who took chocolate milk out of school lunches are promoting puberty blockers, hormone injections, and mutilative surgeries for children who are more than a decade away from having fully developed brains.

It is time we start holding woke public health officials responsible for their dangerous advocacy. It is the only way we'll be able to protect our children from this extreme, irresponsible zealotry.

During the pandemic, the American people learned the importance of questioning anyone who claimed to speak for "the science." They learned that most people who do this are usually pushing a political agenda and using their "expertise" as a cudgel against anyone who dares to stand against them. These experts know that their credentials and high positions in government mean that it's very difficult to challenge them; they can always claim that anyone who does is ignorant, radicalized, or unwilling to accept the settled science. Thankfully, that illusion is beginning to dissipate as public health officials are being held to higher standards.

But as we've seen, this is a problem that goes far beyond public health. Over the past few decades, technocrats have embedded themselves in every aspect of American life—the economy, public works, and even our education system.

The Purpose of Politics

In the modern Republican Party, it's easy to assume that more government is always the wrong answer. Reading this chapter, you might have gotten the impression that I always agree with that assumption. But I don't. While I do believe that the government in the United States has become far too big and not accountable enough to the voters—especially when it relies on unelected experts to make decisions, rather than representatives who were elevated to power by the people—I do see a role for good government. Any sensible elected official should.

Since the founding of this nation, the United States of America has become the most prosperous nation because of deliberate decisions made by people in government—decisions that were made by statesmen and a virtuous citizenry that participated, sacrificed, and cooperated with one another. Some of the most dramatic improvements in technology and life came precisely because of early government investment, further developed by innovative companies that saw a market opportunity to make life better in America. What is alarming about the last thirty years is that our policymakers have forgotten many of the critical policy lessons that made America strong. Our business leaders have stopped caring about America.

And our citizenry has become jaded about participation, divided against itself, and—especially among our elites—more concerned about their own personal success than the success of the nation.

When people disengage, it gives room for the experts to run wild. It sets up a permission structure for the experts, who have spent their entire lives studying one narrow field, to begin making decisions that have vast implications far beyond their field of study and with real impacts on how people truly lead their lives.

To these "smart people," politics is a dirty word. It is a messy, chaotic process that does not lead to good outcomes. Why trust the popular opinion of the unwashed masses when people with multiple Ivy League degrees and countless peer-reviewed journal articles can take the wheel? Between their expertise and the wisdom of the market, nothing can go wrong. After all, "history is over."

But history is not over, and we need politics. To engage successfully in politics, we need a theory of the case. We need to be able to articulate what we will do with power once we get it.

The left has no trouble answering this question. Their party has been taken over by Marxists who have a clear vision of exactly what they will do when they get power. They will tear down the institutions that have made America strong. They will teach our children that the United States is a racist place, and that the traditional path to the American Dream—working hard, living responsibly, and passing along a better future to one's children—is nothing more than a form of white supremacy. Rather than supporting communities and governing by the will of the people, they will continue to allow groups of unelected experts complete control of our civilization.

On the right, we tend to fall back on a free-market fundamentalism—one that views government decision-making as inherently dirty and corrupt, preferring to let the market decide everything.

This view cannot continue. There is a better theory of the case—namely that America is the best nation the world has ever seen, and it is worth defending for future generations. We didn't earn that status because of the size of our economy, the glitz of Hollywood, or the occasional eye-popping returns produced by Wall Street. We earned that distinction because our nation was founded on the fundamental truth that all men are created equal and endowed by their Creator with certain unalienable rights. And over time, we've made strides toward that more perfect union promised by our founders. All of the great struggles we learn about in school—the end of slavery, women's suffrage, the civil rights movement—were led by Americans calling on America to become more like the America our Founding Fathers envisioned.

We have sought to give everyone the opportunity to be raised in a good family and be molded into a good citizen. To work. To engage in community. To raise a family and have the ability to give that family the gift of security and opportunity. To one day leave his children better off than he started and retire in dignity.

In places like Russia and China, where autocracy rather than democracy rules, leaders claim that representational government is difficult and unnecessary. They believe that you can't govern a large nation in the way that we do—that their mode of governance, which allows for five-year plans and government by dictatorial edicts, offers greater stability, security, and prosperity. They tell their citizens that freedom isn't necessary because a government run by experts will provide for the masses. If that sounds familiar, it should. It is exactly the pitch Fauci and his band of merry bureaucrat thugs made during Covid.

We cannot prove these people correct. This is our moment to engage in politics with a true vision of how America remains great, and to prove America's critics wrong.

Chapter 6

AMERICAN NATIONALISM

In the summer of 1960, after about four years of living and working in the United States, my father boarded a ferry bound for Cuba with my brother Mario. During their time in Miami, my parents had managed to make a good life for themselves. They both worked for a time at the Hialeah factories, and my father had earned enough money to afford a new car. In fact, they were only taking the ferry (as opposed to a flight) so my father could show the car to his brothers back home.

But they missed Cuba. Like most immigrants, their life in the United States was not yet full of the same rituals that attended their old lives. They missed their family, their friends, and their culture. You have to remember, it would be decades before the city became a Cuban American stronghold. At the time, Miami was essentially a small, very seasonal southern city. In the late 1950s my grandfather had returned home to Cuba, which deepened my parents' sense of isolation. So when they began hearing that the dictatorship of Fulgencio Batista had been toppled by young revolutionaries in 1959, they made plans to return home and see if things were different in Cuba. My father and Mario went first, and my mother and sister flew over a few days later to meet them.

According to American newspapers, there was reason to be hopeful. A few years earlier, Fidel Castro had visited the United States and assured the public that he was not a Communist. Even as he nationalized oil companies and took other troubling steps toward authoritarianism, he maintained this stance in public.

That hope was shattered when they arrived because my father's brother Emilio saw through Castro's facade. During the weeks that my parents remained in Cuba, Uncle Emilio showed them the early warnings of what the young Marxist revolutionary would soon become. He told them about political dissidents who'd been imprisoned, newspapers and radio stations that had been taken over by the state, and the troubling rhetoric that was coming from the capital.

When that trip was over, my parents returned to the United States for good. Other than a single trip my mother made in the early 1960s to care for her injured father, they did not return. My father never saw Emilio or his other brothers again. In the years to come, they fought through the hardship and loneliness that come with the immigrant experience and put down roots in the United States, eventually having more children and moving to cities all over the country in search of the American Dream.

I've always been grateful that they did. At the time, conditions were even worse under the surface of Cuban society than even Uncle Emilio had realized—especially for kids. Shortly after coming to power, the Castro regime had turned to schools as the primary means of spreading the ideology of communism. Children, in the words of the revolutionary leader Che Guevara, a man Castro had met and admired immensely, were "malleable clay from which the new person [could] be built with none of the old defects."

In 1960 the Castro regime had established state-run nursery schools and a "children's auxiliary club, known as the pioneers, to instill in children between the ages of six and fourteen a love of country and revolution." The next year, he nationalized schools and began designing a literacy program that would turn even the teaching of basic phonics into a tool for instilling the values of Lenin and Marx in the children of Cuba.

This is the program that Bernie Sanders, speaking about five decades later on CNN, would praise for its efficacy in raising Cuba's literacy rate. And it bears a striking resemblance to current efforts to change curricula in America, with critical race theory being the most obvious example—an attempt to mold a new version of American history and identity into the next generation. This is incredibly dangerous to a nation. Fortunately, the vast majority of parents understand this, which is why it has become such a potent issue in American politics. Parents don't want their kids taught how terrible America is. They want their kids to learn to read and do math, the basic skills necessary for a good life. Given that children will have spent sixteen thousand hours in school by the time they turn eighteen, it will also be where they learn values like treating others as you hope to be treated and the dignity that comes from working hard and achieving something. These are values first taught in the home—at least, ideally—and then reinforced in school, at church, and in sports.

Among parents in Cuba, there was a concern that the state would soon attempt to replace them. Many parents worried that their children would be "taken away . . . and shipped to the Soviet Union for indoctrination." Others worried—quite rightly, as it would turn out—that children might someday be forced to report to the government if they heard their parents saying negative things about the revolution at home. Just a few years later, this was common practice in Cuba.

Beginning in 1960, some of these parents did what might have seemed unthinkable just a few years earlier. Working with the Catholic Church, they packed their children's things, gave them special papers from the US government, and put them on planes bound for Miami. Operation Peter Pan, as it is now known, marked the beginning of a massive change for the city. It also provides important lessons for today when it comes to the value of faith, community, and patriotism.

While some of these children had relatives in Florida or other areas of the United States, the vast majority did not. The US government partnered with Catholic Charities of Miami to care for them once they arrived. The children longed to see their parents again—and many were eventually reunited—but the church saw to it they learned to speak English, learned about America and its culture, and were prepared to succeed in their new life. And the success stories abound, a testament to America's unique ability to make dreams come true through hard work, faith, and community. Armando Codina, Max Alvarez, and many others—these incredible individuals all left an imprint not only on Miami but on the nation as a whole. And that was part of the government's plan: to support those in Cuba opposing Castro's rise by protecting their children, but also ensuring that America remained a beacon of hope and freedom to the world in the midst of an ideological struggle between good and evil. By the end of 1962, more than fourteen thousand children had come over as part of Operation Peter Pan.

Over the course of the 1960s, more than a quarter million Cubans fled Castro and came to the United States. The Miami Freedom Tower, much like New York City's Ellis Island, became the location where the government processed these exiles and they began their new lives in America. It was a largely orderly process done under the rule of law—a far cry from our immigration process today. I chose to announce my run for president of the United States from that very building.

For my parents and the many others who came over during those years, America was a place that had offered them an escape from poverty, subjugation, and political persecution. It was a place where they would be able to advance their station in life simply by showing up to work every day and contributing to their communities. Over the course of several years, as they continued to work and put down roots in their neighborhoods, they became true Americans. As their dreams of returning to their homeland faded, they became stakeholders in the well-being of America.

For most of this country's history, that story has been the predominant one. People leave their homes in search of a better life and come to the United States to live here. They get jobs, buy homes, and put down roots. Eventually, most of them became just as American as any of their neighbors who'd been born here. They moved into more diverse neighborhoods, married people from other cultures, and began merging their own traditions with those of the people around them.

In doing so, they learned a fundamental truth about the United States of America: that we do not find our identity in a shared racial background. In other countries, this is not the case. Japanese people share a common ancestral history; the same is true of Chinese people, Germans, and citizens of just about every country in the world other than this one.

The United States, on the other hand, has the unique distinction of being the only country in history that was founded on the basis of ideas rather than a shared ethnic or religious heritage. America is a place, with a distinct culture and set of values, but it's also a nation uniquely able to absorb and integrate the societal contributions of anyone who champions those values, no matter his or her origins. Those ideas are familiar to anyone who's ever attended elementary school in this country and read the opening lines of our Declaration of Independence. Unlike the Communists in Cuba, who were always fighting for a revolution that was immoral and wrongheaded, we don't have to design grammar lessons that sneak these values in through the back door. They are, in the words of our Declaration of Independence, "self-evident."

In our schools, children learn that America has always been a nation of immigrants. The immigrant story, from the lines of Irish and Italian families who lined up at Ellis Island to the thousands of Cuban children who came over during Operation Peter Pan, resonates with us to this day. Contrary to what some on the political left would have you believe, the immigration system in the United States is one of the most open and welcoming in the world. My family's story is a perfect example of that.

When this system works properly, it serves as an example to the world. Unfortunately, it has not worked for some time now.

"My Only Loyalty Is to the Immigrant Community."

Today, that immigration system has been corrupted and exploited. It began, as many of America's problems do, with the fundamental shift toward a globalized economy.

For decades now, since our elites began to believe that global market integration was an inherent good that should take precedence over all else—particularly patriotism, national unity, and the welfare of American workers—this country has prioritized the importation of cheap labor. The 1990s were dominated by stories of local factories shutting down in Pennsylvania and Ohio, only to reopen weeks later in Mexico or China. Offshoring American jobs made headlines and prompted outcry from politicians, but otherwise continued apace even as our communities were hollowed out.

But not every business could be exported, which meant Wall Street simply figured out how to import cheap labor, much of it coming from illegal immigrants. This was a slower, more subtle process. Sure, some politicians made a big deal about "jobs Americans wouldn't do," but otherwise the only outcry came from workers who found their wages stalled, benefits cut, and hours slashed until they could be replaced by someone willing to work more hours for less. More often than not, it is about jobs Wall Street doesn't want Americans to do because hiring Americans would require higher wages and better working conditions. To them, it is better to import cheap labor and buy off Americans with cash welfare programs provided by the government.

What's more, many sought to rationalize the entire process in glowing, nonexploitative terms. They were convinced that the end of history made us all "citizens of the world." Why should jobs in America be reserved for Americans if borders don't matter? In a global economy, the source of the labor doesn't matter, so long as the job gets done. That policy consensus accelerated the erosion of national identity and patriotism.

By the end of 2001, shortly after the United States had helped

China become a member of the WTO, elites from both parties were willing to support the idea that our borders should effectively be thrown open. In the wake of agreements like NAFTA, the world was now "flat," as Thomas Friedman of the *New York Times* would put it, and the conventional wisdom was that our government should act accordingly. In those days, anything that was good for the free market was perceived as being good for the United States, too. No one in our government seemed to spend much time considering the possible complications that would come from making policy this way.

And unlike today, these calls for open borders did not come only from the political left. For years the editorial board of the *Wall Street Journal* had been calling for an open immigration system. In one memorable editorial from 1984, they called for a new amendment to the Constitution reading "There shall be open borders." When Mexico elected a president who called for open borders in 2001, the *Wall Street Journal* editorial board called him a "visionary," and suggested that opening our borders completely was a good idea that "might not unleash a new flood of immigrants."

It wasn't only newspapers. Perhaps one of the most surprising conversations to come out of free trade globalism was America's organized labor movement. Sure, they continued to advocate "Buy American" protections and tariffs, but they increasingly embraced unchecked immigration.

"In a significant policy shift," the *Washington Post* wrote in 2000, "organized labor today called for amnesty for an estimated 6 million illegal immigrants and repealing current law that imposes sanctions on employers that hire them."

The business community was thrilled. "It's a welcome embrace of amnesty from an employer's perspective," the US Chamber of Commerce said at the time. And why wouldn't it be? If near-slave-labor wages could be paid in the United States, they didn't really need the hassle of moving production to China or Mexico. Of course, that happened anyway. Just as organized labor prepared to see jobs shipped overseas, it welcomed a flood of illegal immigrants. It was truly stunning. No one was fighting for the American worker.

By the time Barack Obama began his campaign for president, amnesty for illegal immigrants and open borders were the de facto positions of the Democrat Party. American workers were right to wonder who was fighting for them.

In the years since, Democrats' embrace of cheap imported labor continued, but something equally corrosive began to emerge. Consider, for instance, the reception that Representative Luis Gutierrez of Illinois received when he declared, rather memorably, that he had "only one loyalty, and that's to the immigrant community." This came in the midst of his fight with President Obama over immigration in 2010. By then, the problem of illegal immigration was growing worse by the day, and Rep. Gutierrez had become the leading voice on the issue in Congress. When President Obama first took office in 2009, he had promised Rep. Gutierrez that he would work to pass the DREAM Act. But in the two years since, Rep. Gutierrez had come to believe that President Obama was dragging his feet on the issue, so he went on the attack. He accused the president of focusing too much on apprehensions and deportations at the border, and not enough on granting citizenship to people who were already here.

As a member of Congress, Gutierrez swore an oath to "bear true faith and allegiance" to the Constitution and the United States. But his declaration of loyalty to those who broke our nation's laws was hailed by the liberal media. During the most heated fighting over immigration, *Newsweek* published an article calling him "as close as the Latino community has to a Martin Luther King figure."

At the time, I was living in Miami-Dade County, an incredibly diverse place. I was speaking with members of the "immigrant community" that Rep. Gutierrez claims to represent every day, the vast majority of whom came here legally and followed the rules. Repeatedly, I heard complaints from these people about the swaths of illegal immigrants who were marching toward our southern border and expecting automatic citizenship. Yet whenever anyone in Congress raised those same objections, they were referred to—usually by Rep. Gutierrez and people like him—as racist, xenophobic, or intolerant.

In truth, there are two immigrant communities in America. The one that gets the most attention in our national debate is the group that wants to ignore our laws and change our traditions. This is why so much of our attention is spent on granting amnesty to those illegally here. Activists like Rep. Gutierrez and the legacy media tend to ignore or dismiss the other group, which very much resembles past immigrant communities—people who fled oppression, came legally to America looking for opportunity, and want to be American. We need to stop accepting the idea that the first group is the dominant group. That group would radically change America, not just the Democrat Party.

For this new class of illegal immigrants, the United States was simply a means to an end. It was not something they were part of. Even though many of them were hardworking people, America was just a place they could live for a while and earn money to send home to their families who were struggling back home. From a human standpoint, this impulse is fully understandable. But in most cases, these weren't people who felt that they were tied to America in the long term. They weren't stakeholders in the well-being of this country. Their ability to be successful wasn't tied to their embrace of America, our communities, or our traditions. As a result, many immigrant communities became more insulated, and ethnonationalism among them became more pronounced. It has become possible, for instance, to drive down an American street and see the flags of a dozen different countries, not a single one of which was the United States. It became common for children to grow up without speaking English, or to go through school without learning the history of the country to which they had immigrated.

While some commentators have suggested that assimilation has become more difficult in the past thirty years, that is not the case. I'm from one of the most diverse cities in the United States, and I see assimilation all the time—every time I walk into a Cuban restaurant and see the dishes of my childhood served with a Colombian side, or fried yucca served at an American pub. You see it at a high school football game—a uniquely American sport embraced by everyone.

But I've also seen people who seem dedicated to stopping assimilation in its tracks, particularly in Washington, DC. Usually, these are people from the left-wing activist class whose political careers depend on keeping immigrant communities angry, isolated, and resentful. The dumb word "Latinx" is a great example. The first time I heard it, I thought it was patronizing and absurd. I don't know a single person outside of Washington, DC, who uses it. Too often, journalists and pollsters make the mistake of assuming that these people represent the views of all immigrant communities.

They don't.

Amnesty by Another Name

As Obama's time in the White House came to a close, the divide between the liberal activist class and the actual immigrant community became a chasm. The language that was used to attack those who called for serious immigration reform grew more detailed and more intense. It was also used far more frequently. During the 2016 campaign, anyone who put forward a policy that was not complete and total amnesty was equated with a fascist or a militant white supremacist. If a working-class white person who'd been forced out of the labor market by the influx of cheap, illegal labor raised concerns about the prospect of allowing even *more* foreign workers to come into the country illegally, that person was treated with the kind of contempt that we once reserved for members of the Ku Klux Klan.

Even the smallest offenses provoked outrage among the ranks of the Democrat Party, which was shifting further to the left with each passing moment. When Bernie Sanders, an avowed socialist who was in the middle of running the most far-left campaign in the history of American politics, suggested that it was not a good idea to elect someone purely on the basis of their Latino identity, a representative from the Hillary Clinton campaign suggested that he "may be a white supremacist, too."

During President Trump's four years in the White House, the left doubled down on this kind of language. Suddenly, "white supremacy" was the defining feature of the United States, and Donald Trump—who tapped into real concerns about rising immigration rates—was the ultimate white supremacist. Anyone who defended him or attempted to work with him on immigration was, therefore, probably a white supremacist, too. In their eyes, this meant that continuing to split people up based on their identity was not only permissible but necessary.

In 2019, after years of protests against Immigrations and Customs Enforcement, or ICE, immigration activists in Colorado infiltrated the agency's headquarters and pulled down the American flag. They also removed a Blue Lives Matter flag, covering it with spray paint and scrawling the words "Abolish ICE" on it. In place of the American flag, they raised a Mexican one. While their tactics were extreme, these activists were not at all out of step with the modern Democrat Party. During her successful campaign for Congress, Alexandria Ocasio-Cortez had repeatedly called for the abolition of ICE. So had Representative Mark Pocan of Wisconsin and several other representatives.

Speaking in 2018, Senator Elizabeth Warren said that we needed to "rebuild our immigration system from top to bottom." Anyone who looks at our border or knows someone who has gone through the process legally would find themselves in agreement with Sen. Warren's observation. Our system is clearly broken. Not only have we lost control of our border, but we've also lost control of who comes legally. What do I mean by that? Well, who comes legally is essentially a lottery ticket. You either get selected randomly for "diversity" reasons, you're lucky enough to have your family sponsor you, or you can buy your way in. There is no grand plan to make sure our immigration system is focused on building a stronger America.

But the rest of Sen. Warren's comment makes clear she doesn't care about any of that. Instead, she wants to rebuild our immigration "starting by replacing ICE with something that reflects our values." Last year, US Customs and Immigration Enforcement seized record volumes of fentanyl and meth coming across our southern border. The agency also apprehended a record number of suspected terrorists crossing into our country. What would Sen. Warren have it do instead? It was just a cheap, throwaway line crafted to cater to the Democrats' increasingly radical base. Just like the calls to "defund the police" that would come two years later, the statements reflected a growing propensity toward lawlessness in the Democrat Party. It also revealed how dependent our nation's elites had become on the politics of resentment and racial division.

Years earlier, before the Democrat Party became ruled by an unhinged group of anti-American zealots, there was an opportunity to rebuild our immigration system to reflect our values. It came at a moment when the Democrat Party was changing for the worse. Barack Obama's most radical impulses were unrestrained after winning reelection, and he had no interest in protecting our nation's borders. There were plenty of Democrats in the Senate who felt otherwise, or at least that is what they said.

What became known as the Gang of Eight bill would have radically curtailed family-based chain migration, which would become a huge priority for President Trump years later. It would have dramatically increased border security, including $46 billion for border patrol, fencing, and more. It would have required E-Verify for employers and biometric data from immigrants. And we would have tied our immigration system to merit—to make sure that people coming to our nation actually benefited America, just like every other nation approaches immigration.

While it wasn't popular at the time, I was willing to take the political risk because the window for securing our border was rapidly closing. Democrats were radicalizing. And if those security provisions were enacted, Trump would have had nearly unlimited resources just a few years later to finally secure our border. But as those negotiations continued, it was increasingly clear that Democrats were not actually interested in any of that. Instead, their primary goal was amnesty for anyone and everyone who had already come to our country illegally. Every draft resulted in weaker enforcement language, weaker trigger language, and more administrative flexibility for a lawless administration. Rather than serving as a starting point, the effort imploded as Obama continued to do everything possible to undermine trust in the rule of law.

As it turned out, that was the last time Democrats would even bother to pretend border security was important. Just look at what happened in 2018, when President Trump offered to grant legal status to children brought into the country illegally by their parents in exchange for only $5 billion in border wall funding and minor changes to family-based chain migration. Democrats responded with cries of racism and shut down the federal government to block Trump's proposal. Less than a year later Democrats would again shut down the government, this time for thirty-five days—the longest shutdown in our nation's history—again to block border security funding.

I've learned the hard way that the national Democrat Party simply has no interest in being an honest partner in negotiating immigration laws that make sense and can be—and will be—enforced.

And we are now seeing the logical conclusion of their decade of obstruction. Our borders are wide open, and the Biden administration is granting a de facto amnesty to millions every year through nonenforcement of the law.

In May 2022, US Customs and Border Patrol apprehended more than 239,000 people attempting to cross our southern border. The last time the number was that high was March 2000, which happened to be right when those labor unions called for amnesty.

The reason for this latest sharp increase is simple and predictable: Joe Biden invited them in.

During the presidential campaign of 2020, Joe Biden sent a clear message to the world. Unlike President Trump, who had made immigration enforcement a key element of his time in the White House, Biden signaled that he would do the opposite. During campaign speeches, he pledged to shut down all detention centers along the border. He promised that no illegal aliens would be deported during his first hundred days in office, and he insisted repeatedly that the United States was not accepting enough immigrants.

Speaking in a conference room in Iowa, Biden said that the United States could "take in a heartbeat another two million people. The idea that a country of 330 million people cannot absorb people who are in desperate need and who are justifiably fleeing oppression is absolutely bizarre. Absolutely bizarre. I would also move to increase the total number of immigrants who could come to the United States."

Think about the absurdity of that statement for a second. Seven hundred million people live on less than $2 per day. Every single man, woman, and child would benefit from coming to the United States. Where would Biden, Harris, and organized labor draw the line? They have opened the floodgates and lost control of the border.

The results were predictable. When Joe Biden took office in January 2021, the number of illegal immigrants encountered by Border Patrol was hovering between 70,000 and 80,000 per month. One month later, Border Patrol encountered more than 100,000 illegal immigrants. That March, the number shot up to 173,277, then to 178,795 in April. One year later, in April 2022, the figure was an astounding 234,088. While it was the highest number of monthly border encounters in twenty-two years, it quickly became the Biden-era norm.

According to photos published in *Business Insider*, many of these wore T-shirts reading "Biden, Please Let Us In!" as they attempted to cross from Tijuana.

Like millions of immigrants before them, these people were

coming to the United States for a better life. But unlike those millions of past immigrants, they did not believe that they needed to follow any kind of rules before crossing our borders and attempting to settle in this country. I'm sure many of them had heard stories from friends and family who'd made the journey before them about the various changes to our immigration system that occurred as soon as Joe Biden took office. They knew that while there was a good chance that they would be detained and deported, there was an equally good chance that they'd be given a court date, placed on a bus, and shipped to the interior of the United States, where they could successfully evade the government for the rest of their lives. For many of them, that was a chance worth taking.

Unlike many of my colleagues in the Senate, I know these people. I know their families. Many of them live in South Florida. When I speak to them, I hear stories about the terror that they fled Honduras, Guatemala, and other countries to escape. I hear about the horrible economic conditions that many families are forced to contend with in their home countries, and the stories about neighborhoods that are completely controlled by street gangs. I hear about fathers who are being extorted by these street gangs, and mothers who are afraid to send their children out to school for fear that they won't come back.

At the same time, it simply cannot be the case that any time something doesn't go well in some Latin American country, people all of a sudden have a right to immigrate to America. Refugee and hardship immigration can be a complicated moral issue, but we cannot deny the fact that often people should *stay* in their home countries, or nearby, and work to fix them. That is not always the right solution, as my family's history demonstrates, and when an entire country is overtaken by communism, as in Cuba or Venezuela, it can be wise for the United States to extend grace to immigrants. But that is not the same thing as a right to immigrate, and the ultimate solution, for Cuba as elsewhere in America's backyard, is to promote good self-government, whatever it takes.

Still, as a father myself, particularly one whose own parents fled a country they loved, I understand why these people make the decision to leave. If my own children were ever in the kind of danger that I've heard about from recent Central American immigrants, there is nothing I wouldn't do to get them out of that danger immediately.

But the Biden administration is not doing these people any favors by pretending that the United States will be able to give them immediate asylum as soon as they cross our borders. Like any country in the world, we have immigration laws, and we need to enforce them. This means, among other things, that everyone who applies for asylum needs to be vetted and processed in an orderly manner. Back when the numbers of people attempting to cross our border illegally were relatively low, processing all of them was already a difficult task. Now that Joe Biden has effectively lifted all restrictions on border crossings, it's become impossible.

As a result, the routes that these people take to enter the United States in secret are among the most dangerous in the world. Men, women, and children regularly die on the journey. In the summer of 2022, Border Patrol authorities in Texas came upon a tractor-trailer truck with sixty-four migrants packed into the back with no air conditioning. By the time they arrived, forty-eight of these people had already perished in the heat. Tragically, this is not an uncommon occurrence along our southern border. It's been happening for decades, and it'll happen much more often if we don't take steps as a country to address the root causes of this crisis.

One of the most important steps, as I have said for years, is to strengthen the economies of the countries these people are fleeing. This is why I have always been a strong supporter of the Alliance for Prosperity, a program designed to strengthen the economies of El Salvador, Guatemala, and Honduras so that people will no longer feel the need to leave those countries for the United States. As I've pointed out, we do not have a migratory crisis from Costa Rica. We don't have a migratory crisis from Panama, Chile, or Peru, because these are countries where people can live without fear of crime and lawlessness.

Most importantly, though, we need to enforce our existing immigration laws and make sure the rest of the world knows that we are willing to do so. For too long, we have sent the opposite message, and the consequences have been devastating. The Biden administration spent years fighting desperately to end the pandemic-era Title 42 authority that allows for swift removal of migrants caught entering our country illegally. Even under Biden, more than one million people have been swiftly removed with this authority. Last year, El Paso mayor Oscar Leeser—a Democrat—declared a state of emergency ahead of the expected expiration of Title 42. He understood that Title 42 was one of the few tools remaining in the Biden administration's arsenal to deter migrants. Democrats responded by spending billions of dollars to help process illegal immigrants more quickly into the country.

Our immigration system, which has always been one of the most open and accepting in the world, has become overwhelmed by the sheer numbers of people who believe they can come in and take advantage of our generosity.

Admittedly, this mistaken impression did not come out of thin air.

With every statement like Biden's, the United States saw a further erosion of national identity and patriotism. That statement says the only requirement to come to America is that you want to come to America. Nothing about wanting to become an American—to learn our nation's values and language, to work hard, live responsibly, and participate positively in this great nation's civic life. More and more, children whose parents came over the border in the late 1990s and early 2000s were raised in communities that were insulated from the rest of the United States—communities in which a refusal to assimilate was seen as a good thing. In time, this intense division became the default position of many immigrant communities, especially illegal ones.

In previous generations, it was the children who broke this cycle. While the parents and grandparents of recent immigrants were usually reluctant to assimilate, still speaking their native language at home and retaining most of the customs of their home country, children who attended school in the United States and made friends of different races and creeds would usually blend in with their surroundings just fine. They would bring some of the traditions from their parents' culture, such as food or music or literature, while also embracing the other traditions that make up American society.

But this is becoming less and less common. In a world where one of our two major political parties continues to insist that we divide people based on their race alone, assimilation becomes not only difficult but socially unacceptable. Today, young immigrants are growing up in a world where there are hundreds of different "affinity groups" to join on college campuses, where they are constantly reminded of their "diversity" by Democrats, and where they often feel as if the American Dream is out of their reach.

The results, particularly for young people, have been devastating.

Indoctrination

In October 2015, a student at Claremont McKenna College wrote an essay in her school newspaper. The student, whose parents were working-class immigrants from Mexico, wrote that she felt "out of place" on her college campus. According to her essay, this is because the college—and the country, for that matter—makes people from immigrant families such as hers feel "abnormal" and "unwelcome."

In the first paragraph, the student tells a story about being made to feel shame that her father, a first-generation immigrant, made his living as a waiter.

"When my fifth-grade teacher asked the class to write about our career of choice," she writes, "I, of course, wrote 'waiter.' He politely asked if I could choose something else."

As the son of a bartender, I found it hard not to feel a certain amount of empathy.

But it dissipated quickly as I kept reading.

"One of the many things I've learned from queer activists," the student writes, "is that assimilation does not equal liberation. Achieving the 'American Dream' for myself does not mean that people like my parents, relatives, or hometown community will stop being dehumanized or that they will be given the respect they deserve. . . . These feelings [of inadequacy] caught me by surprise as I had never known what it felt like to be the 'minority' in my predominantly immigrant, low-income Latinx hometown."

In these few sentences, you can find almost everything that has gone wrong in the debate over immigration and inclusion in this country. When I was a young man, I often felt the same lack of belonging that this student writes about. But back then, the answer for young immigrants was not splitting off into racial affinity groups, tearing at our imagined wounds, and penning dramatic op-eds in student newspapers about our irreparable sense of unbelonging. We did not siphon ourselves off from American society and create new, nonsensical terms like *Latinx* to distinguish ourselves.

Instead, we got to work and absorbed the culture around us—and we contributed to it. We assimilated because it was the norm—it seemed natural. When I was growing up in the 1980s, the political left was focused on labor unions and helping members of the working class. When they did embrace identity politics, they did it for the cause of *inclusion* rather than exclusion, arguing that everyone should be granted the same rights no matter their ethnic heritage. In doing so, they followed the example of great civil rights leaders such as Martin Luther King Jr., who once called our founding documents "a promissory note to which every American was to fall heir." In other words, our mission was to bring people together under one shared national identity.

But slowly, beginning around the 1990s, elite institutions began adopting identity politics as a tool for exclusion and division. Activists spoke of interlocking oppressions and the impossibility of understanding anyone else's life story. In the years that followed, we were introduced to completely new terms: safe spaces, toxic whiteness, and "microaggressions," the latter of which has been defined by a professor from Columbia University as "brief and commonplace daily verbal, behavioral, or environmental indignities, whether intentional or unintentional, that communicate hostile, derogatory, or negative racial slights and insults toward people of color."

In 2014, just a year before this student at Claremont McKenna published her essay, the University of California published a list of supposed microaggressions. One of the terms was "melting pot." Apparently, when a "person of color"—another coinage of woke activists—heard the phrase "melting pot," all they heard was a white supremacist telling them that they didn't belong in the country. Obviously, this was nonsense, and it should have been treated as such by intelligent people everywhere, especially on college campuses.

But it wasn't. For a while, anyone who dared speak out against this deranged ideology was quickly silenced for fear of being "canceled." People were fired. Others were threatened with the loss of their livelihoods. Most of the time, this happened to people who had no ill intent. It happened to people who had simply chosen the wrong words, used a phrase they shouldn't have, or forgotten to include some piece of language that the mob demanded of them.

A few days after the student at Claremont McKenna published her essay, for instance, a woman named Mary Spellman reached out to the author via email. For years, Spellman had been the dean of students at Claremont McKenna, and it seemed that she wanted to let the student know that her concerns, unhinged though they might have been, hadn't fallen on deaf ears.

In her email, Dean Spellman thanked the student for writing, invited the student to speak with her, and stressed the importance of serving students "who don't fit our CMC mold."

Now, if most people had received that email, they probably would have taken their dean of students up on the offer. That's certainly what someone who was interested in changing things for the better might have done. If they weren't interested, they might have ignored the email or deleted it.

Instead, the author of the essay took a screenshot of the email and posted it to her Facebook page about two weeks after she'd received it. In a message attached to the image, she mocked Dean Spellman for saying she "didn't fit the CMC mold," and asked her friends to share the image.

Which they did. A few days later, according to an account of the event written by Jonathan Haidt and Peter Lukianoff in their book *The Coddling of the American Mind,*

> *the campus erupted in protest. There were marches, demonstrations, demands given to the president for mandatory diversity training, and demands that Spellman resign. Two students went on a hunger strike, vowing that they would not eat until Spellman was gone. In one scene, which you can watch on YouTube, students formed a circle and spent over an hour airing their grievances—through bullhorns—against Spellman and other administrators who were there in the circle to listen. Spellman apologized for her email being "poorly worded" and told the crowd that her "intention was to affirm the feelings and experiences expressed in the article and to provide support." But the students did not accept her apology. At one point a woman berated the dean (to cheers from the students) for "falling asleep" during the proceedings, which the woman interpreted as an act of disrespect. But it is clear from the video of the confrontation that Spellman was not falling asleep; she was trying to hold back tears.*

In the end, Mary Spellman resigned. The students who'd berated her for racism celebrated. The whole thing was over in a matter of weeks.

By the time it happened, the sequence of events was so familiar that it hardly warranted a mention in any major newspapers. It seemed that every few weeks another person on a college campus was fired or forced to resign over increasingly silly imagined acts of racism.

Underlying all these incidents was a strange new set of rules about how immigrant groups should be treated. According to these new rules, the color of a person's skin is of paramount importance. If a white person says something, it is interpreted differently than if a "person of color" says the same thing.

For years, members of the elite commentator class had been ignoring the troubling evidence of student protests, claiming that once these misguided children had to go out into the world and get real jobs, they would come to their senses and abandon these strange, neo-Marxist ideas.

But that didn't happen.

Instead, they carried this anti-American ideology into newsrooms, boardrooms, and tech companies all over the country. They got jobs in classrooms, teaching our children that the United States of America is an evil, racist place built on a foundation of white supremacy. They began teaching white children that they were complicit in this white supremacist system and that they were guilty for the country's "original sin" of slavery, while telling black children that they would be oppressed by this racist system forever.

All the while, these same people suggest that we should keep our borders open and allow everyone in the world access to this vile, racist society. The contradiction does not seem to bother them. And if it does, they certainly don't show it.

To them, immigrants of this country all hold the same opinions as extremists such as Luis Gutierrez. Anytime a member of the immigrant community speaks out against the policies of the Democrat Party, the representatives of that party assume it's because of "misinformation." As Biden's first White House press secretary, Jen Psaki, has noted, the left believes that immigrants are particularly vulnerable to misinformation. That, in their eyes, is why the Republican Party has done so well with Latino voters lately.

Again, nothing could be further from the truth. If anything, the sudden upsurge in support for Republicans among recent immigrants is a sign that they are seeing things more clearly than ever.

Echoes of the Past

For years, I've found that the most patriotic people I know are the ones who've come to the United States from another country—particularly if that country had a repressive Communist government, the way Cuba did when my parents came here in the late 1950s. They came to America for the same reason our founders did.

Anyone can learn about the difference between the American system of government and the government in, say, Venezuela. They can read the newspaper or pay attention during the nightly news. But only someone who has recently arrived here from Venezuela—who has seen firsthand how socialism turned an oil-rich country into one of the most backward economies in the world, who has been forced to carry stacks of hundred-dollar bills to afford food because of rampant inflation—can truly appreciate the difference.

More than once, I've seen the dedication that recent immigrants have to this country. A few years ago, I was sitting in the waiting room of a doctor's office near my house with about a dozen other people, almost all of whom were Cuban immigrants of various ages. After a few minutes, one of the people who'd recently arrived began complaining about how bad America was. He claimed that here, unlike in Cuba, he wasn't given enough by the government. He complained that he had to work too hard.

The response from the older people in the room was immediate. Before the young man could finish speaking, several people in the room were yelling in his direction. They told him how lucky he was to be here rather than back in Cuba. *If you don't like it,* they said, *then leave.*

I've seen the same scenario play out everywhere: bars, restaurants, even grocery stores. For years, Cuban people have been coming to grocery stores in the United States and loading up long duffel bags with consumer goods to ship home to their families in Cuba. Everyone knows what these bags, called *gusano* after the Spanish word for "worm," are for. If you look at any flight that's preparing to leave for Cuba, you'll see a big pile of them getting loaded onto the plane.

I vividly remember standing in line at the supermarket recently and watching a young man load up one of these bags at the register. Between me and him were about a half dozen older Cuban immigrants, all of whom waited patiently as the young man finished bagging his items. When it came time to pay, though, he pulled a food stamp card out of his wallet.

Again, the condemnation from the older folks in line was swift and loud.

It was understandable. These were people who had worked for thirty or forty years to earn enough money to live, and most of them had probably retired with nothing more than the value of their house and a social security check. The man in front of them, on the other hand, had manicured nails and the latest iPhone, and he was paying for goods to take home to his relatives with government benefits. For people who grew up in an environment where people expected handouts from the government and then *left* that system in search of a society where people are able to make their own way in life, this was particularly offensive.

The examples go on and on. There is a phenomenon in South Florida—although it is not exclusive to that area—whereby wealthy couples from Latin America will fly into the United States when their baby is about to be born. This way, their child will be born an American citizen, and will therefore be entitled to all the benefits that citizens enjoy. Often, they will leave without paying their hospital bill.

Under Joe Biden, who has invited illegal immigration with his policies, this will only get worse.

In fact, it already has.

At the end of 2022, this country began seeing a surge of migrants arriving in the Florida Keys, mainly from Cuba and Haiti. Between December 30 and January 5, more than eleven hundred Cubans and Haitians poured in, overwhelming national resources and resulting in the mobilization of the Florida National Guard.

Once again, some of the people who were most upset about this surge were not anchors on Fox News or wealthy Republican donors. They were Hispanic men and women who came here legally and got jobs to support their families, never relying on the government for support. Over breakfast in early January, I could hear a woman from Central America talking to her waitress about all the benefits that the people who came here illegally would now get; her tone was much the same as the older Cubans who watched the young man pay for his items with a food stamp card.

It is not an accident that the loudest cries against socialism usually come from the Latino community. Unlike the college students who read a few pages of Karl Marx and think Bernie Sanders has some good ideas, people who've lived under Communist dictatorships know where socialist policies lead. They understand that socialism will always fail because it views wealth as immoral, and that the people who push it are usually doing so because they want more control over the lives of citizens.

When Cuban immigrants hear, for instance, about the various far-left lessons that are now commonly taught in American schools, their thoughts return immediately to where they came from. When they hear stories about children who've been told to go directly to their teachers if they feel they need to change their gender rather than confiding in their own parents, they are reminded of the exhortations they got as children to report on their own parents for not being supportive enough of the revolution.

Above all, these people know that they chose to come to America for a reason—the same one that brings so many people rushing toward our borders every year. They know that this is by far the most welcoming country in the world, and that our economy gives people the chance to rise through the ranks of society like nowhere else on earth.

Chapter 7

ANTI-AMERICA, INC.

"What's Good for General Motors . . ."

In 1953, shortly after he was elected in a landslide, President Dwight D. Eisenhower began building his cabinet. For most positions, he selected men from the usual places—government, academia, and the US military, an institution he had famously led during the worst fighting of World War II and its aftermath. But when it came time to name his secretary of defense, a job of immense importance at the time, given how fragile the incipient global order was, Eisenhower turned to the business community.

Specifically, he wanted a man named Charles E. Wilson, who had served as president of General Motors for about a decade. This wasn't as strange as it might have seemed at first. During the war, Wilson had overseen GM's efforts to manufacture key military equipment for our soldiers overseas. According to a recent report, Wilson was able to turn GM into the "largest military contractor on earth" during the war, "making 119,562,000 shells, 206,000 aircraft engines, 97,000 bombers, 301,000 aircraft propellers, 198,000 diesel engines, 1,900,000 machine guns, and 854,000 military trucks."[19] Previously, while working at the Westinghouse Electric Company in Pittsburgh, Wilson had helped develop equipment for the army and navy during World War I. With Wilson at its helm, the Department of Defense would be run by a man with experience running vast bureaucracies and, perhaps most importantly, who knew the importance of a strong industrial base to the national defense.

In those days, it was not uncommon for private companies such as General Motors to work for the US government, nor was it uncommon for these companies to view themselves as existing primarily for the benefit of the United States and its people. That is why, when war broke out twice, it was our major American corporations who snapped into action and began making the supplies our military would need, powering this country on its way to becoming an economic and military powerhouse by the end of World War II. The norm of our society then was that the leaders of these companies had an obligation to consider the good of the nation when making major business decisions, and, for the most part, they did.

In most cases, I'm sure it didn't even occur to them that we might someday live in a world where the leaders of American companies did *not* consider their businesses to have an obligation to serve their nation.

This notion was brought into sharp relief in January 1953, when Charles Wilson appeared before the Senate Armed Services Committee for a two-day confirmation hearing. At issue, primarily, was the fact that Wilson still held stock in General Motors that was worth about $2.5 million, which was an enormous sum of money at the time. Adjusted for inflation, the stocks would be worth about $27 million today—not enough to crack the top twenty in today's Senate, but a substantial amount nonetheless.

Throughout the hearing, several senators wondered aloud whether Wilson's stock holdings might create a conflict of interest. Not only did General Motors still hold several contracts with the Defense Department, but the company would almost certainly do more business with the government in the future. The way those deals were handled might cause GM's stock price to rise and fall, which would affect the net worth of a sitting defense secretary. After several hours of dancing around the issue, one senator finally asked the question plainly.

"I am interested to know," he said, "whether if a situation did arise where you had to make a decision which was extremely adverse to the interests of your stock and General Motors Corp. or any of these other companies, or extremely adverse to the company, in the interest of the United States government, could you make that decision?"

According to the *Congressional Record,* Wilson answered in the affirmative: "Yes, sir; I could. I cannot conceive of one because for years I thought what was good for our country was good for General Motors, and vice versa. The difference did not exist. Our company is too big. It goes with the welfare of the country. Our contribution to the nation is quite considerable."[20]

Almost immediately, a version of this quote began floating around Washington, and then the country. People came to believe that Wilson had uttered the line—now famous—that goes, "What is good for General Motors is good for the country, and what is good for the country is good for General Motors." Despite attempts to clear things up in the next week's *New York Times,* which said, according to the Library of Congress, that "some of those present [at the hearing] do not recall that Mr. Wilson made the widely quoted remark," the phrase soon entered the popular lexicon.

Although Wilson ultimately did sell his stock in General Motors, his notion that the interests of GM and America were, effectively, one and the same was a good definition of that era—one in which American companies hired American workers, made their products in America, and reinvested their profits for the benefit of the American working class. Looking back, his statement (which he never *quite* said) might seem arrogant, and his insistence that he be allowed to retain his stock might seem misguided by today's standards. But there is no doubt that this American adage was true, at least for most of the twentieth century.

From the postwar decades forward, US corporations helped to make our country the most prosperous in the history of the world. They gave American citizens the means to rise above their stations, offering stable jobs and dignified work for fair pay. In turn, they reaped enormous profits. But with those profits came a corporate duty to care for the strength of the nation and its citizens.

Today, that bargain has broken down. Indeed, it is difficult to imagine the CEO of a company such as Apple or Nike going before Congress and claiming that "the difference did not exist" between what was good for their companies and what was good for America. If anything, the opposite is often the case. What is often good for the short-term interest of these companies ends up being harmful to the long-term interests of our country. When Apple began manufacturing many of its products in China, for instance, that was undoubtedly bad for America. By going to China, Apple gave the ambitious and aggressive Chinese Communist Party access to some of the most advanced telecommunications technology in the world. Communist China didn't have to buy, steal, or develop it; instead, Apple simply gave them access. In turn, China gave Apple access to cheap labor and hundreds of millions of new consumers. It was a win-win for Apple. Besides, why pay American wages to build iPhones "designed in California" when slave-labor wages are just across the ocean?

There is no denying that the move was good for Apple, at least in the short term. Over a period of a few decades, the company was able to become the first corporation in history to achieve a valuation of a trillion dollars, something that was virtually unimaginable just a few years ago.

At some point over the past thirty years or so, corporate America began to view families, communities, good jobs for Americans, and even the nation itself as afterthoughts. American workers of all backgrounds suffered as a result. It is not an exaggeration to say that the hollowing out of corporate investment, outlined in the second chapter of this book, has annihilated an entire way of life.

Soon after, a culture shift followed. Executives began to espouse cosmopolitan values, downplaying American patriotism. Love of country, free speech, traditional faith, and other bedrock American ideals became unfashionable. And suddenly some American companies—in many cases, the same ones that had mobilized to ensure the safety and security of our soldiers during World War II—stopped viewing our national security as a priority.

In 2016, for instance, after a mass shooting at an office building in San Bernardino, California, carried out by two jihadists named Syed Rizwan Farook and Tashfeen Malik, the FBI announced it was unable to unlock an iPhone that belonged to one of the shooters. In an attempt to find out whether this person might have been connected to any other potential terrorist threats, the FBI asked Apple to bypass the passcode on the phone so that federal agents could access the information within. Given that the two shooters had pledged allegiance to ISIS and other online terrorist groups, their concerns were certainly well-founded. But Apple refused, claiming that the code necessary to break the encryption of the iPhone would be "too dangerous to create."

After a lengthy battle in our court system, a magistrate judge ordered Apple to "assist in circumventing the encryption and accessing the iPhone's content." But Apple, continuing to insist that this would constitute a serious violation of privacy and personal security, did not follow the order. Instead, Tim Cook, who'd been the CEO of Apple for a few years at the time, published a document titled "Message to Our Customers," which described the FBI's efforts to unlock the shooter's phone as "chilling," and suggested that Apple felt compelled to "speak up in the face of what we see as an overreach by the U.S. government."[21]

As the writer Yishai Schwartz has pointed out in *National Affairs*, it is notable that "Tim Cook's open letter announcing Apple's decision to fight the FBI and the federal court order was not a letter to the American public; it was a 'message to *our customers*'—American or otherwise."

Schwartz continues:

> *The company has sheltered billions of profits in tax havens overseas, putting it in similar standing with 30 of the country's largest multinational corporations. And in a fascinating interview with the*

Washington Post, Cook revealingly responded to a question about his responsibility "to publicly take on such issues as civil rights" with his own discourse on human rights. The difference between the two terms is subtle but significant. It reflects the difference between a worldview that places the nation—and the rights that citizenship affords—at the center, and one in which moral concern and loyalty are diffuse and globalized.

Of course, Apple isn't too concerned with "human rights" either, at least not when profits are at stake. The company infamously made changes to its App Store and AirDrop sharing features to appease the Chinese Communist Party in the midst of mass protests. Only a few years earlier, Cook had announced Apple's $100 million Racial Equity and Justice Initiative, tweeting, "The unfinished work of racial justice and equality call us all to account."

Apple is not alone in the sudden shift from "corporate patriotism" to pandering for profit. In 2013, during a dinner at the White House, Facebook founder Mark Zuckerberg—who was, at the time, desperate to expand the reach of his company into the Chinese market, wanting to fully embrace his role as a "citizen of the world"—asked Chinese general secretary Xi Jinping to give his unborn child a Chinese name. This practice, which is common in China, would forever link Xi to Zuckerberg's child; the relationship would be almost familial. Ultimately, Xi declined, reportedly to avoid the solemn responsibility that comes with giving a child a Chinese name.

While Zuckerberg now likes to brag that his company is not doing business in China, his tactics indicate he's shifted ever closer to those of the Chinese Communist Party. In the years since he was rebuffed by Xi at the White House, Zuckerberg and other leaders at Facebook have brought their company much further along the road to authoritarianism, adopting speech codes that seem to come right out of the Chinese Communist Party's handbook. Every day, Facebook and Instagram—both owned by the megacompany now known as Meta—censor any speech that Meta's (liberal) employees deem to be misleading, hateful, or unworthy of representation in the public discourse. In recent years, the Biden White House appears to have directed these companies to remove any posts that contradict the administration's guidance on Covid-19 policy—which, as we've seen, is often proven to be false or misguided within days of being released.

But it's not just large tech companies. The problem of corporate leaders putting political correctness and short-term financial gain over their obligations to our country afflicts most of corporate America.

Stakeholder Capitalism

Today, the very same companies that outsource American jobs to China and elsewhere and rely on slave labor also flex their power to humiliate Americans who dare to support any traditional values at all.

Multinational firms threaten boycotts over pro-life legislation. When the Supreme Court handed down its decision in *Dobbs v. Jackson Women's Health Organization*, which simply stated there was no constitutional right to murder an unborn American, several major companies sent out notices guaranteeing that if any of their employees still wanted to get an abortion, regardless of what state they lived in, the company would cover the cost of travel out of state, and in some cases, of the procedure itself. While few of these companies used the word *abortion*, opting for less inflammatory terms such as *women's health care*, it was clear to anyone reading these notices that the corporations did not agree with the Supreme Court's ruling, and that they were attempting to circumvent it using their considerable money and influence.

When the state of Georgia passed a series of commonsense voting laws—which included a requirement that voters show identification at the polls, clarified rules for drop boxes, and expanded mandatory early voting—the uproar in corporate America was even more intense. Almost immediately, Major League Baseball announced that it would move the annual All-Star Game out of Atlanta, saying that the "MLB fundamentally supports voting rights for all Americans and opposes restrictions at the ballot box."[22] Delta Airlines—a company that makes billions every year in large part through deals with Communist China—said that the bill was "unacceptable and [did] not match Delta's values."

A few days earlier, more than a hundred CEOs of major companies had participated in a Zoom call organized by Jeffrey Sonnenfeld, a professor at Yale University School of Management, which trains some of the top business leaders in the United States. By the end of this call, which included the heads of companies such as Walmart, Levi Strauss, LinkedIn, and others, the CEOs signed a letter indicating "readiness to act individually and collectively to shore up American democracy and ensure Americans have access to a world-class voting system." Those present, according to the *New York Post*, also "indicated they will re-evaluate donations to candidates supporting bills that restrict voting rights and many would reconsider investments in states which act upon such proposals."[23]

Every few days, it seems, the corporations of this country take another far-left political stand—one that we are assured is brave, progressive, and necessary given the times we are living in.

Some of this is undoubtedly motivated by greed. The leaders of these corporations want to sell more products, and they believe that one way to achieve that goal is by parroting whatever woke nonsense is popular at the time. And it is certainly easier for corporations to change their marketing to cater to progressive social tastes than it is to do the hard work of developing an innovative product against competition. Some of it is also driven by the many corporations' increasingly progressive employee base, as young employees go on to staff corporations after being educated in leftist ideologies on their college campuses.

But corporations have gone woke even in the face of serious

opposition from regular Americans. According to a recent survey by American Compass, 63 percent of nonmanagement workers want businesses to "focus on business and stay out of social justice issues" like "election reform, racial equity, and LGBTQ+ rights." Among the workforce generally, those numbers went up to 66 percent of independents and 85 percent of Republicans.

It didn't happen overnight that nearly every corporation in the Fortune 500 became crushingly progressive against the wishes of the American public. And it goes against common sense: Why would corporations do something that undermines their existence and creates opposition to them over the long run? There is a deeper cause.

As a senator, I talk often with corporate executives about the economy and how their businesses can help to advance the national interest. Though many corporate executives are good people who want to do the right thing, patriotic corporate leadership is too often in the minority. Today, too much of corporate leadership—including, importantly, some of the biggest investors on Wall Street—would rather cave to leftist social activists in their workforces, on social media, and even in the boardroom.

Corporate America has gone woke ultimately because corporate America's sense of duty has shifted away from our nation and its people. It first shifted, with the rise of globalization, away from a duty to the nation to a duty of maximizing financial values devoid of long-term investment. Once these bonds of nationality were dissolved, corporate America developed into a new kind of class. Corporate leaders wanted to maintain their good standing with other members of the global financial elite and became susceptible to the intellectual fads of that class. When wokeness took over, it was, to use Professor Sonnenfeld's phrase, like a "spiritual awakening." No longer subject to duties to invest their immense resources for the good of the country, large corporations became instruments of the political left.

This shift tracks with the evolution of the main trade association for large corporations, the Business Roundtable. In 2019 this group of top executives made headlines when they issued a new statement on "the purpose of a corporation." Since 1997, the group's guidance had stated that corporations should exist only to advance the interests of their shareholders; everything else, in their view, was peripheral. But according to its new guidance, conceived by J.P. Morgan CEO Jamie Dimon and written by Alex Gorsky of Johnson & Johnson, companies should also consider their *stakeholders*—often code for left-leaning political issues like climate change—when making business decisions, not just shareholders.

It also describes the position of the largest fund managers in the world, like BlackRock. BlackRock, which by 2020 managed nearly $9 trillion, is a leading player in the revolution on Wall Street to push ESG—environmental, social, and governance—initiatives such as reducing carbon emissions and conducting "racial audits" of companies. But this is only BlackRock's latest move. In addition to using "stewardship" to prod companies on leftist social issues, it uses its immense leverage over companies to push for financial returns over capital investment. Take, for example, BlackRock's support for the election of green-energy activists to the board of directors of ExxonMobil. These activists pledged to cut back on oil and gas not only because of their claims about climate change but because doing so would increase financial returns for shareholders. And in the short term, they might be right: drilling is one of the most capital-intensive activities in the world. But it is necessary for both long-term corporate performance and the good of the nation. If a company has nothing to invest in, then it will simply return that money to shareholders. In the dominance of fund managers like BlackRock, we see financialization and corporate wokeness melded into one.

It's important to recognize that all of this started with a movement, starting in the 1980s and '90s, to make stock value the sole criterion of business success. Companies began to realize there was more money to be made in the short term through complex financial instruments and offshoring jobs to foreign countries than there was in funding innovation and product development at home. While that might be sustainable in the short term—might, in fact, be extremely *profitable* in the short term, especially for companies that had cut costs by shipping major operations overseas—it was not a model that could be relied upon to get this country where it needed to go. And everyday Americans knew it.

During the campaign of 2016, I had the privilege of speaking with many people who felt as if they had been left behind by the free-market ideology that had become popular in the Republican Party over the past few decades. Since roughly the 1970s, it seemed that the Republican answer to every problem was to put more faith in the free market, to cut taxes even more, and to trust that American corporations would invest in American citizens and help our economy to thrive. Sometimes it worked out, and sometimes it didn't.

Many of these people had lost their jobs, and they didn't have the option (as many policymakers seemed to think they did) of moving to Silicon Valley and learning to code or design the newest iPhone.

When I returned to the Senate after the campaign, I spent a great deal of time trying to develop economic policies that would support a free market, but also make that market work for the people rather than just the tiny number of Americans whose income is derived primarily from the stock market. I thought of my parents, who were able to support my family on the income of a bartender and a maid for most of my childhood. Although I knew that the world I grew up in was impossible to bring back, I also knew that the world we were currently living in was unsustainable, for the simple fact that it depended on a system that did not value American workers. I devoted a lot of time and energy to understanding this issue and fighting for workers, even when it was unpopular.

For example, during negotiations for the Tax Cuts and Jobs Act of 2017, during which many Republicans argued that lower corporate tax rates would trickle down to communities and families, I attempted to ensure that the bill would directly benefit working families. Some of those provisions, such as expanding the Child Tax Credit, made it into the finished text of the bill.

My guiding principle, as always, was that the free market was useful to the extent that it served the people of this country. As soon as it stopped doing that, some changes needed to be made. But the exact nature of those changes was up for debate. Over a period of about six months, I worked with the best people I could find to fix the problem. I started by meeting with scholars also seeking to understand this problem, such as Oren Cass and Julius Krein, and building out a dedicated team on the US Senate Committee on Small Business and Entrepreneurship to investigate the problem.

In the end, what we found was astounding, though it didn't surprise us. America's economic growth was being driven more by finance than innovation in the production of real assets. And it was not by accident. The report we put out argues that since the 1970s, changes made by American businesses and policymakers began prioritizing high returns to investors in the short term, rather than investment in long-term capabilities. The *Washington Post* called the report an "unsparing rebuke of the business community." It opened a lot of eyes, and probably cost me a few donors, including people who donated to my presidential campaign. But I believe that taking on the monumental issue of corporate governance is one of the great projects for conservatives in our time, and it is worth the cost.

You see, when corporate America started shedding its obligations to the national interest, justified by an economic ideology that privileged stock value over all else, it wasn't going to stop there. Shareholder value is a subjective value that has no necessary connection to the good of the nation. If all of the big funds on Wall Street said they would only invest in companies that, for example, committed to "net zero" carbon emissions, then the stock prices of those companies would likely go up. For example, a 2020 study confirmed the intuitive idea that when a company gets included in a stock index that large investors are all invested in, like the S&P 500, the company's stock price goes up. And the so-called ESG movement has long been clear that its goal is to raise the cost of capital for disfavored companies by denying them access to large banks and institutional investors. A company's shareholders might well include large foreign shareholders with goals contrary to our national interests.

Conservatives need a new understanding of corporate purpose. During the summer of 2019, I began preparing to deliver a speech at the Catholic University of America, one of the most prominent religious schools in the United States. According to the invitation, I could have spoken about anything. We were coming on three years into the Trump presidency at the time, and there were several issues that I believed deserved attention—the rising threat from China, for instance, and the partisan hostility that seemed to be building among voters in the lead-up to our next election. But in the end my team and I decided to use the speech to call for "Common Good Capitalism," an economic model in which companies would care more about making this country and our communities strong than they would about their rising stock price.

In the speech, which drew heavily on the teachings of Pope Leo XIII, I noted that the choice we have been given between unlimited free markets on the right and socialism on the left is a false one. Conservatives, I said, have championed the obligation to work and neglected the rights of workers to share in the benefits they create for their employer, while liberals have championed everyone's right to various benefits while ignoring the obligation to work. In the end, I laid out several policy proposals that my team and I had been working on for months. I suggested revamping existing structures to encourage investment in small businesses, writing tax policy that would discourage stock buybacks, and expanding the federal Child Tax Credit.

The criticism came quickly, primarily from the political right. Some commentators accused me of adopting socialism, which was absurd. As the son of a family who fled a socialist regime for the United States, I am the last person in this country who needs to be lectured about the evils of that economic system.

I am an enormous supporter of free-market capitalism because I've seen and know that it has generated more prosperity than any other economic system in human history. But when that system stops working the way it's supposed to work, it creates structural imbalances that lead to many of the problems we are seeing not only across the United States but across multiple developed Western economies. We are certainly doing well in the short to mid-term, but we also have an obligation to think about the structural challenges that will make it difficult to sustain that growth in the long term.

The Forgotten American Worker

Capitalism is one of the greatest inventions in human history. It brilliantly allocates resources toward the most efficient outcome. But the most efficient outcome is not always what is best for our country. In a business plan, "labor costs" are indistinguishable from "equipment costs," "taxes owed," or "property insurance." Each is a dollar figure that subtracts from what is deemed the most important line item of all: "profit." That may work just fine when it comes to accounting, but it is not the optimal way to run a country. To a nation, workers are not the same as equipment or insurance. Workers are fathers, mothers, neighbors, and citizens.

It is the job of policymakers to ensure that the most efficient allocation decision is also the outcome best for the nation. If it's not, we can change the rules to align the incentives of the private sector with the national good. While the free market should always be the first choice, we must be able to acknowledge when the market does not necessarily know best. But doing so requires a conception of the national good. And that, ultimately, is what our elections and our public debate need to be about.

We cannot be a strong nation without a strong workforce. Dignified work makes family and community possible. These are not financial markers, but rather the markers of a successful and enduring nation. Without dignified work, society decays. Suicides and drug overdoses soar. Values erode. There is no more important project than rebuilding our nation's workforce, and that begins with rethinking our century-old approach to organized labor.

The modern labor movement abandoned any pretense of supporting American workers two decades ago when the AFL-CIO, the nation's largest union, came out in support of amnesty for eleven million illegal immigrants. But the last few years have made clear just how out of step these organizations are with their members. Just look at what happened when the unions representing rail workers sat down with the rail companies to negotiate a new contract. The workers—everyone from traveling linemen to conductors—were desperate for some sort of paid sick leave. Labor leaders negotiated just about everything else, and then told their members to take the deal, which still did not include sick leave. Not surprisingly, the workers balked. If our system worked, the labor leaders would have gone back to the companies and restarted negotiations. Instead, the labor leaders and business executives sat down with the Biden administration to force the agreement on the workers. And Biden, along with a Democrat-controlled House and Senate, did exactly that.

I was one of the few who refused to go along with the charade. To be clear, I don't think Congress had any business involving itself in negotiations between the parties, but if the businesses and labor leaders came crying to us, you can bet I was going to side with the workers. Someone has to, because at the behest of Wall Street, rail companies had dramatically cut the number of workers over the past several years to maximize profit. Fewer workers working more hours with less time off equals more for shareholders. It is the classic example of efficient allocation of resources, except that workers are actual human beings with families. They are members of their local church. They volunteer to coach their kids' soccer and basketball teams. But in the eyes of the neoliberal elites—which includes their own union leadership—they are just an expendable resource.

It was a very clear example of the breakdown of our current labor system and how the neoliberal consensus marginalizes those who do the hard work, but it is far from the only one. Amazon, for example, has a long history of highly questionable labor practices. Just as the railways used something euphemistically called "precision scheduled railroading," Amazon uses a dystopian system called "time off task" to track nearly every minute of employees' days. It is crazy and demeaning, which is why I stood with the workers at Amazon's Bessemer, Alabama, distribution warehouse when they attempted to unionize in 2021.

While I am not a fan of our archaic labor laws, they were the only tool available to the workers in Bessemer. And, equally as important, I surely wasn't going to side with Amazon—a company that increasingly uses its market share to silence conservative voices, such as Ryan T. Anderson, the author of *When Harry Became Sally*, and push an insane woke agenda. It's funny; all these companies love to condemn Republicans, but as soon as they need relief from regulations or protection from unions, they come running to us. They don't get to have it both ways. If they want to call me a bigot because I believe only women should play women's sports, then they shouldn't expect me to run to their rescue, especially when their ask is to side against working-class Americans who are trying to provide for their families.

Thankfully, Republicans are increasingly skeptical of corporate America. It is an important and long-overdue shift. For the most part, the leadership of these major companies have more in common with Democrat elites than they do with the hardworking Americans on their payroll who increasingly vote Republican. But what remains to be seen is whether—and how—my party grapples with the role of organized labor. The truth is that our current system is adversarial by nature. It writes conflict and strife into our legal code. There are times, like we see with Amazon, where conflict is both necessary and inevitable. But there are also times when the current arrangement breaks down and leaves the actual workers without real representation, like it did during the rail negotiations.

We need to be open to supporting traditional organizing where it makes sense, but also to developing a new concept—a twenty-first-century version of labor relations. Instead of making it illegal for employees and management to work through problems together, we should allow for the creation of voluntary employee involvement organizations. The concept is widely used in Germany with great success, and would serve as an alternative to the failed status quo. There will be resistance, of course. Resistance from labor leaders. Resistance from Wall Street. And even resistance from companies. But none of that should deter us from making changes to strengthen our workforce.

I believe that ensuring that our labor markets produce enough jobs, requiring diverse skill sets that match the capabilities and interests of our citizens, is an important part of the national good. Many corporate leaders instead seem to think that lowering labor costs so as to prioritize shareholder returns is their mandate. I believe bringing our supply chains home and reducing our dependence on China is in the national interest; many corporate leaders would rather push woke nonsense to appease global financial elites.

And at the same time, they have used all the terms that the political left tells them to use. They've donated to the right causes, and they've adopted all the right hiring initiatives. To put it bluntly, the leaders of these corporations have paid deference to everyone *except* the American workers they are supposed to serve, and they have attempted to distract us from that fact with empty woke buzzwords. For the most part, they have done this while claiming to support "people of color" in the United States, a group viewed by the political left as being completely oppressed by racism. Yet their solutions to solve this constant barrage of systemic racism—often exaggerated or in some cases completely imagined for political reasons—has not been to ensure, for instance, that black children grow up in more stable households or that their neighborhoods are made safer by more police. Instead, the woke left—and, for some reason, many of America's major corporations—have offered these groups little more than diversity training, new buzzwords, and a complex set of terms to describe their suffering.

But as we moved into 2020, a year when the world experienced a global pandemic and several months of violent rioting in the streets, we learned just how damaging this new mentality could be.

During this tumultuous period, it became clear that many multinational corporations in the United States were not only ambivalent about their obligation to Americans; in many cases, they were downright hostile to this country and our values.

Going Woke

During the summer of 2020, we saw this trend in action. Almost as soon as the first protestors appeared in the streets of Minneapolis, several major American companies released statements of solidarity with the Black Lives Matter movement. Repeatedly, these companies implied that systemic racism was the most serious threat to American citizens, ignoring a great deal of data that proved otherwise.

In the weeks that followed, corporate America seemed to imbibe every dictate of the radical left, hiring diversity consultants and decrying "whiteness"—a term straight out of the critical race theory curriculum—in their corporate correspondence.

Several companies, most notably Coca-Cola, implemented strange courses on "how to be less white" by the writer Robin DiAngelo, whose book *White Fragility* shot up the bestseller list in the aftermath of these protests. While these courses were in use, the employees at Coca-Cola were forced to endure a video that told white employees to "be less oppressive; be less arrogant; be less certain; be less defensive; be less ignorant; be more humble; listen; believe; break with apathy; break with white solidarity."[24]

In a widely shared video taken as our streets were being overrun by riots and looting, Jamie Dimon took a video of himself getting on one knee in his office, an apparent sign that he was on board with the Black Lives Matter movement. Later, he said America's history was one of "systemic racism."

In a sense, this parroting of woke ideology gives these companies cover to keep embracing the policies that have led us to where we are today. Leading financial figures such as Larry Fink, who runs the largest hedge fund in the country, will talk endlessly about the new "stakeholder capitalism." He'll discuss the need to fight racism, protect the environment, and ensure that corporate boards have at least one member of every race, gender, and sexuality. Not only are these issues nonsensical, pushed by a small group of activists using online platforms that are governed, for the most part, by *other* liberal activists; they are distractions from the blatant corporate malfeasance of some of the most vocal supporters of stakeholder capitalism.

All the while, Fink's company BlackRock continues to tout investment opportunities in Communist China, a country that is carrying out a full-scale genocide and is focused on rendering America a second-rate power.

Several other companies have proven that they are more than happy to criticize the United States—the country that has allowed them to grow and prosper for years—while continuing to appease murderous dictatorships like the Chinese Communist Party. Last year, for instance, it was revealed that Warner Bros. censored dialogue that suggested a gay relationship for the Chinese release of the latest film in the Harry Potter franchise. Meanwhile, in the United States, anyone who raises the slightest objection to having LGBTQ themes crammed into every children's movie is deemed a bigot and subjected to cancellation campaigns on social media.

These hypocrites want to have it both ways. They want to coast off everything that makes America the most business-friendly country in the world, while moving jobs out of our nation and waging a merciless war against traditional values.

So far, they've succeeded. Getting in bed with the Chinese Communist Party has opened up enormous new markets. Outsourcing jobs has been a tremendous cost saver. Bending a knee to woke progressive craziness has made CEOs more popular than ever in elite settings.

But increasingly, the bill is coming due. More Americans are realizing what I understand. According to a poll taken by the Brunswick Group, only 36 percent of voters agree that "companies should speak out on social issues," while 63 percent of corporate executives agreed with the same statement. A few months earlier, according to the *New York Post*, 37 percent of adults said that "Coca-Cola's swipe against Georgia's voter integrity law made them less likely to purchase Coke products."

As our corporate leaders care less and less about the strength of our nation, the policy advice they give lawmakers makes less and less sense for our country.

Cutting corporate taxes—and especially investment taxes—makes sense if US companies are going to invest in American industry. But if they are instead prioritizing offshoring corporations or simply returning windfalls to shareholders, then policymakers are going to start being more careful in how we structure tax cuts.

Employer-friendly labor laws make sense in a world where corporate CEOs feel an obligation to their fellow countrymen and workers. But the logic of resisting labor representation on behalf of corporate management falls apart if an American worker is no different from any other input to the corporation.

Taking aggressive positions on woke cultural issues that tear at our national fabric might seem like an easy way to avoid boycotts from activists. But those of us charged with keeping America strong recognize that these positions are the greatest threat to our long-term viability.

No policymaker would allow a company to dump toxic waste into a river upstream of a thriving town he is charged with governing. Yet corporate America eagerly dumps toxic woke nonsense into our culture, and it's only gotten more destructive with time. These campaigns will be met with the same strength that any other polluter should expect.

Our nation needs a thriving private economy. And patriotic business leadership has historically underwritten the American Dream. But lawmakers who have been asleep at the wheel for too long, especially within my own party, need to wake up. America's laws should keep our nation's corporations firmly ordered to our national common good.

The work of rebuilding—and rebalancing—the relationship between our nation and its large corporations begins today.

Chapter 8

THE LIBERAL CULTURE WAR

Written Off

During his 2008 campaign, at a fundraising dinner in San Francisco given when he was still a junior senator from Illinois, Barack Obama delivered a line that would soon become famous. On its face, his speech was about the problems facing this country. Ideally, he was supposed to tell the room how he planned to fix those problems.

A few minutes into his speech, Obama began describing a few things he had seen on the campaign trail. The images he described were the same ones that I would see a few years later when I campaigned for the presidency, often in the same cities and towns that Obama toured in 2008.

"You go into these small towns in Pennsylvania," he said, "and like a lot of small towns in the Midwest, the jobs have been gone now for twenty-five years and nothing's replaced them. And they fell through the Clinton administration, and the Bush administration, and each successive administration has somehow said these communities are gonna regenerate and they have not."

Looking back, you might be slightly surprised at how perceptive these words seem, especially considering the policies Obama would come to adopt once he became president. Back in 2008, the Democrat Party still tried to appeal to working-class voters—many of whom were the backbone of the historical Democrat coalition. Its donors, operatives, and intellectuals were all cosmopolitan elites isolated in coastal enclaves, but the vast majority of its voters were hardworking Americans. The problem for Obama at that moment was squaring the working-class coalition that he hoped would elect him with the roomful of liberal elites he was speaking to, whose policies had made so many American jobs disappear in the first place. There were policies he could have pursued as president that would have spoken to a working-class coalition. He might have addressed the globalist policies that allowed so many of our companies to ship key operations overseas. He might also have criticized the trade deals that were sending so many good American jobs to China and Mexico. He might have noted the decline of marriage and of church attendance. But all of these positions were anathema to the neoliberal consensus among the types of people—sadly, of both parties—who donate to campaigns, run campaigns, and staff White Houses.

The comments that I have focused on President Obama making that night, thus far, are not the famous part of his remarks. As we will see, they are unrecognizable in today's Democrat Party, but they are familiar to the Democrat Party that existed for my entire life up until that point. It was a party that I disagreed with on the size of government, the need for a strong American military, and the morality of important issues like abortion. But it was a normal party.

The more famous remarks that Obama made that night are the ones the elites Obama was speaking to really wanted to hear, and they are instructive of where the Democrat Party has gone since 2008. They didn't want to hear about the valid grievances that working-class Americans had about the impact of outsourcing, financialization, and hyperindividualized morality on their lives. America's ruling class hadn't failed the working class; the working class had failed America's ruling class. Speaking about the people who'd been left behind when their companies shipped their jobs overseas, Obama said it was "not surprising" that these people "get bitter; they cling to guns or religion or antipathy toward people who aren't like them or anti-immigrant sentiment or anti-trade sentiment as a way to explain their frustrations."

In these few lines, you can find everything that would come to define the assertive, unpatriotic, anti-American liberal elite for years to come. It was the quintessential liberal definition of the "other." It revealed what the liberal elite truly thought about the rest of America, and represented the fundamental turn in America that happened, sometime after Obama's reelection, that made cultural warfare against regular Americans the defining source of conflict in our politics.

Rather than tying the animosity of so many working-class Americans have toward our political system to the economic anxiety he had just talked about, Obama blamed it on cultural issues. And not just a legitimate disagreement between fellow citizens on culture issues. Worse, Obama suggested—as thousands of left-wing politicians would suggest in his wake—that these people were "bitter," or that they were expressing "antipathy toward people who aren't like them." In other words, these voters only felt the way they felt because they were backward, racist people. To the liberal mind, therefore, we shouldn't feel too bad about leaving them behind with our policies.

Obama helped draw a clear line between liberal elites and the rest of America by declaring that there's something inherently wrong with guns or religion. To the liberal elite, "cling[ing] to guns or religion" is a cultural mark of backwardness. Drug overdoses, suicides, joblessness—all these are self-inflicted, and thus not the fault of America's elite financiers, celebrities, or titans of exported industry.

Senator Obama's off-the-cuff remarks are revealing of a Democrat Party caught between the economic anxiety of working-class voters and an elite consensus that didn't believe there was anything left to debate on economic policy other than minor details. The real fight, in the minds of our nation's ruling class in 2008, was on cultural issues. While there were debates on the size of government, both parties had fully bought into the neoliberal consensus. It was simply a matter of execution. And Democrats rightly perceived Republican weakness on protecting and defending traditional American values. At the time—and even to this day—the most popular position among ruling-class elites was to be "fiscally conservative, culturally liberal." This was how Republican donors said that they agreed with Democrats on the need to move beyond the bitter voters clinging to their guns and religion, while pretending to have profound economic disagreements. In reality, the elites in both parties had more in common culturally with one another than the rest of America. They were eager—in some cases, desperate—to move beyond the bitter voters clinging to their guns and religion.

Fast-forward eight years to Hillary Clinton, the Democrat Party nominee for president, making a similar gaffe while speaking to her elite donors about America's working class. This time, she used a new word, "deplorables," which would soon become far more famous than Obama's "guns and religion" gaffe of two election cycles before.

"You know," she said, "just to be grossly generalistic, you could put half of Trump's supporters into what I call the basket of deplorables. Right? The racist, sexist, homophobic, xenophobic, Islamophobic—you name it. And unfortunately, there are people like that. And he has lifted them up."

By now, it seems redundant to point out all the various ways in which statements like this are repugnant. Most obviously, they violate a basic rule of electoral politics, as several people pointed out during the fallout after Clinton's remarks first appeared on the internet—namely, that while you can insult your opponent all you want, it is never advisable to insult the American voters or their beliefs. In the event that you win, these are the people whose support you'll need, so it's generally considered bad politics to call them "deplorable."

Then, more importantly, such remarks are disgusting from a moral standpoint. As the Lord says in the book of Psalms, "Whoever slanders his neighbor I will destroy."

In the time between Obama's speech and Hillary Clinton's, something changed within our national conversation. Whereas President Obama had been forced to hide his contempt for America's working-class values beneath a recognition of economic trouble, Hillary Clinton felt perfectly comfortable coming right out and saying that a large portion of the American electorate was no longer worth speaking to. This was no longer just a means of identifying the "other"; if you were "culturally backward," you were an enemy.

Clinton, like so many other proponents of the anti-American liberal elite that President Obama represented, seemed to have taken his reelection in 2012 as an indication that the left had won the culture war, and that politics was now about putting the losers in their rightful place.

What changed in those eight years? If you ask many Democrats, the answer is that white America had a racist reaction to President Obama—America's first black president. Jamelle Bouie—currently a *New York Times* opinion columnist but at the time the chief political correspondent for a mediocre ruling-class web magazine called *Slate*—quotes Robin DiAngelo, whom I introduced in a previous chapter as the author of *White Fragility*: "'White Fragility is a state in which even a minimum amount of racial stress becomes intolerable, triggering a range of defensive moves,' she writes. 'These moves include the outward display of emotions such as anger, fear, and guilt, and behaviors such as argumentation, silence, and leaving the stress-inducing situation. These behaviors, in turn, function to reinstate white racial equilibrium.'"

White fragility, according to Bouie, is why Republicans nominated Donald Trump. "DiAngelo was describing private behavior in the context of workplace diversity training, but her diagnosis holds insight for politics," Bouie wrote. "You can read the rise of Obama and the projected future of a majority nonwhite America as a racial stress that produced a reaction from a number of white Americans—and forced them into a defensive crouch."

This, however, is nonsense. When President Obama came into office, he enjoyed a 69 percent approval rating, according to Gallup. Americans were proud of the racial progress our country had made—even if many of us were deeply disappointed that their new president was a committed liberal who would embark on an aggressive left-wing policy agenda. Obama had defeated my Senate colleague, John McCain, in that 2008 election. McCain's concession speech gracefully addressed the pride many of us felt in having America's first black president: "A century ago, President Theodore Roosevelt's invitation of Booker T. Washington to visit—to dine at the White House—was taken as an outrage in many quarters. America today is a world away from the cruel and prideful bigotry of that time. There is no better evidence of this than the election of an African American to the presidency of the United States. Let there be no reason now for any American to fail to cherish their citizenship in this, the greatest nation on Earth."

I believe what actually changed from 2008 to 2016 had little to do with the feelings of the Republican Party and a lot more to do with the feelings of the Democrat Party leadership. It had to do with a feeling that liberalism had triumphed. The culture fights of the past had been won—everybody the Democrats knew understood that guns were evil, religion was disproven by science, and cultural conservatism was an impediment to the higher goal of maximizing individual happiness. These questions were settled. All that was left to be resolved was how to deal with the backward people who hadn't yet gotten the message.

The Liberal Enlightenment

When President Obama first became president, the expectations for him were extremely high. In part, this was because of the historic milestone that his election represented. America had elected a black man to the presidency.

If nothing else, it should have proven once and for all that America was not a racist country.

But it didn't take long for the enthusiasm to wear off. During his first few years in office, President Obama did almost nothing to help hardworking Americans who were struggling—the very people he was talking about during his infamous "guns and religion" address. Instead, he presided over the consolidation of the liberal elite behind a set of issues that would come to dominate our politics today.

Obama's evolution on social issues is a great example. When it came to certain issues, the President Obama who governed during his first term would be almost unrecognizable to the young liberals who revere him today. In fact, if he made some of the statements he made then today—just a decade after he made them—he'd be canceled immediately. During his first term, President Obama was willing to say only that he was "wrestling with the issue" of gay marriage, not that the practice should effectively be made legal in all fifty states overnight (which, astoundingly, is exactly what would happen during his second term). And just years before, the future president in his memoir noted his discomfort with immigrant displays of pride, writing, "When I see Mexican flags waved at pro-immigration demonstrations, I sometimes feel a flush of patriotic resentment. When I'm forced to use a translator to communicate with the guy fixing my car, I feel a certain frustration."

When it came to some of these social views, sometimes President Obama said that his opposition was religious, and that his Christian faith would not allow him to support such policies; other times, he attempted to argue his position from a legal standpoint, claiming, for example, that he did not believe marriage was a civil right to which gay couples could legally argue they were entitled.

Obama's reelection campaign was bitter. It converted the goodwill that accompanied his election in 2008 into a trap. Opponents of his reelection were branded opponents of progress. False allegations of racism ran rampant—much more so than in 2008. Then vice president Biden told black voters that Republicans wanted to put them "back in chains." While Obama's election in 2008 was supposed to represent the end of identity politics, his reelection campaign used the power of the presidency to weaponize identity politics. And it worked.

President Obama, of course, won the 2012 election. This was the moment many of America's leading progressives had been looking forward to for years—decades, in some cases. Now that Obama had been reelected after presiding over two massive expansions of government, these liberal elites came to believe that they had been vindicated in their worldview. Not only did they believe that the culture war at home was over, and that key issues such as gay marriage and abortion rights would soon be solved in their favor; they also believed that their view of domestic politics and foreign policy had been endorsed by the American voter.

Moving forward, the US government would act not as a nation with its own interests whose citizens needed to be protected, encouraged, and supported, but as one more citizen in a growing global community. Rather than shoring up our supply chains and creating good jobs at home, our government would begin to worry more about solving climate change and newly developed theories of systemic racism. At home the Obama administration would enforce these policies through massive expansions of government, which would give it even more power to push through elements of its far-left globalist agenda. As we saw during the campaign, there was virtually no facet of American life that the Obama administration and its compatriots did not believe could be made better with the introduction of more government bureaucrats: health care, education, finance, business, and more.

Most disturbingly, the liberal elites who had cheered on Obama's victory that night would soon come to believe that the president's second victory was the result of a changing America—one that was getting "browner," as the writer Ronald Brownstein would put it in *National Journal*, and younger every year. The theory, which might sound familiar to modern audiences, was that as the United States took in more immigrants—especially Hispanic immigrants, who had begun arriving on our shores in record numbers—and churned out more young, college-educated white women, the voter base of the country would drift steadily leftward with each passing year. Given this "coalition of the ascendant," as this loosely assembled group was called, most commentators assumed that the Republican Party would soon die out if it didn't adapt to become younger, more accepting, and more inclusive of immigrants.

Oddly enough, this notion was never called out as racist for the way that it assumed that all minorities must think alike; no one ever wrote scathing editorials accusing Democrats of pushing "great replacement theory" when they suggested that a new group of immigrants and other minorities was going to usher in a new age of Democrat dominance that would last for the next few decades. These journalists and professors certainly didn't consider the notion that when Hispanic immigrants arrive, in many cases after months of running from socialist dictatorships in their home countries, they would drift not toward the Democrat Party but rather the one that had always supported smaller government and more personal freedom, and warned against sudden radical social change imposed from above.

Back then, in the aftermath of Obama's second victory, that idea would have been considered insane. But as recent polling has shown, it's exactly what has happened. In the years since Obama won a second term, the immigrants who were supposed to bring about sweeping demographic and political change have in fact become swing voters. In the 2018 midterm elections, according to Pew, Hispanic voters backed Democrat candidates by a nearly 40-point margin; in the 2020 election, President Trump won the support of about one in three Latino voters nationwide. According to CNN's exit polls, that number increased to 39 percent in 2022. In my race, I earned 56 percent of the Hispanic vote.

The reasons for this shift are clear, and they have *been* clear for years. But too many of our political pundits and journalists were willing to believe the lie that after President Obama was reelected in 2012, the nation was about to become more liberal, less ideologically diverse, and more receptive to the kind of rapid government expansion that the Obama administration had attempted during its first four years in office. If you were a devoted reader of mainstream newspapers at the time, or even a viewer of the nightly news, you might have developed the mistaken impression that the constant battle between the two major political parties of this country—the battle that brings about compromise, which is the lifeblood of our political system—had ended, just as the Cold War had ended in the early 1990s. And just as at the end of the Cold War, proponents of the new liberal world order had embarked on a historic, unilateral campaign to remake the United States of America in their progressive, borderline Marxist image.

Meanwhile, the top experts in my own party, many of whom were members of the same globalist elite class as the higher-ups in the Obama administration, seemed too willing to accept this grand narrative of progress and sweeping demographic change. Many of them took the political press's story about the "coalition of the ascendant" at face value, not even bothering to question the assumptions underneath it. Almost immediately after Mitt Romney delivered his concession speech, the Republican National Committee began doing research on an "autopsy report," attempting to figure out what had gone wrong during the campaign of 2012 and make suggestions for how the party could run a better campaign next time.

Those suggestions, unsurprisingly, would turn out to be totally wrong in almost every way possible. This is because they were based on the same naive beliefs that would lead the Democrats to embark on their mission of social change and rapid government expansion. Even the top Republicans in the country seemed to believe that as the country grew more ethnically diverse in the years following Obama's second term, it would become dramatically more left-wing as well. They assumed that the only people in the country who weren't thrilled about this sudden demographic change were the old white men who were being phased out, and that the reason for their dissatisfaction was racism. In this, at least, they shared common ground with President Obama, who had famously accused those people losing their jobs at record rates of channeling that frustration into "anti-immigrant sentiment," "anti-trade sentiment," and "antipathy toward people who aren't like them."

In just four years, the belief that the Republican Party could only serve racist old white men had moved from the fringes of the radical left into the halls of the Republican National Committee, an organization that would make its recommendations for the future based on that assumption. It was wrong then, and it is wrong now. But it would occupy the top brass of the Republican Party for the next four years of President Obama's second term. Meanwhile, while the experts in my own party were attempting to find out why immigrants don't like us, the Obama administration began trying to implement its radical plan for America.

To this day, we are living with the damage it did.

An Overnight Evolution

In July 2014, faced with historic (and well-deserved) opposition from Congress on several pieces of his agenda, President Obama gave an outdoor speech in Washington. For the past two years he had been battling with Republican members of the House and the Senate—me included—who wanted to roll back some of the more outrageous expansions of government that his administration had attempted during his first term. These fights were prolonged and often ended in stalemates, but that, as we all know, is part of the process. The American system of government is based on the assumption that the president should not be able to take unilateral action without the approval of Congress, as any child who has taken a basic civics course can tell you.

But President Obama and his administration, still believing that they had the approval of the country to enact a sweeping liberal revolution, began to believe otherwise. Speaking about Congress during his speech, Obama defended his right to act alone rather than trying to work with Congress.

"As long as they insist on taking no action whatsoever that will help anybody," he said, "I'm going to keep taking actions on my own that can help the middle class. . . . Middle-class families can't wait for Republicans in Congress to do stuff. So sue me. As long as they're doing nothing, I'm not going to apologize for trying to do something." The message was clear: rank-and-file voters do not matter, at least not unless they agree with elite liberal opinions. People forget that as recently as 2008 voters in California—obviously no one's idea of a conservative state—voted to define marriage as between a man and a woman. But today gay marriage is legal nationwide, and the dissenters are labeled bigots. The nearly overnight imposition of gay marriage on the entire country by five unelected justices was a seminal moment in understanding the triumphalism the left felt during the Obama years, and understanding how their demands would become so much more in the years that followed.

During an interview with ABC News in May 2012, after years of hinting about the progress he was making on the issue, President Obama announced that he did in fact support gay marriage. "At a certain point," he said, "I've just concluded that for me personally, it is important for me to go ahead and affirm that I think same-sex couples should be able to get married."

Over the course of the next three years, the Obama administration moved unilaterally to ensure that the nation would "evolve" on the issue of gay marriage at the same blinding speed that the president and vice president had. Anyone who expressed even mild skepticism about whether this was a good idea was quickly labeled a bigot or homophobic for daring to suggest that states had the right to define marriage as a union between a man and a woman—something that President Obama himself had apparently believed until just a few years earlier. The Obama administration was triumphalist in its celebration of the Supreme Court's earth-shattering decision. Millions of Americans were deeply shocked by this decision, but their feelings didn't matter to the ascendent coalition that viewed them as backward. The White House was lit up in rainbow colors, and the work of moving on to the next cultural fight began. Few Americans in 2015 would have believed some of the fights we see today over sexual orientation and gender identity. You would have been called a conspiracy theorist if you said that a male would win the NCAA women's swimming championship in 2022, or that a battered women's shelter would be forced to accept biological males.

In the fight over gay marriage—not that there really *was* a fight at all—we can see a familiar pattern emerge: first, activists on the radical left lay down a hard-and-fast rule about a group that now, according to them, deserves a special constitutional right that did not previously exist; next, they lobby the president to announce support for that rule; then, finally, they label anyone who doesn't get on board immediately as a bigot, a racist, or some other woke phrase that carries the threat of immediate cancellation.

When I first entered national politics, such rapid change would have been difficult. During the early 2000s, for instance, outrage traveled mostly by handwritten mail and telephone lines to radio call-in shows. You'd receive a handful of emails, but only half of adults had internet access at the time, and clickbait wasn't yet a thing. It took months, sometimes years, for a critical mass to develop and force action. However, with the advent of social media, social change came incredibly quickly. In some cases, this was undoubtedly a good thing. The elites no longer controlled the flow of information. However, the elites quickly learned how to weaponize the decentralization of information in the manner described above. Issues that had previously been discussed only on the furthest fringes of the radical left were now mainstream, and anyone who didn't agree was deemed unworthy of participation in our political process.

Many people, lawmakers and Supreme Court justices among them, warned that the so-called progress that the left was making would not stop with gay marriage. Soon, they warned, the ever-growing acronym that described people who deserved special protection (just LGBT at the time) would grow to encompass more and more groups, nearly all of whom would claim greater levels of imagined oppression.

They were right.

In December 2014 the Obama administration took key steps to ensure that the Civil Rights Act of 1964 would protect not only Americans of all different races and genders but also those who were transgender. Schools, according to a memo from the Department of Education under Obama, "must treat transgender students consistent with their gender identity in all aspects of the planning, implementation, enrollment, operation, and evaluation of single-sex classes." Speaking shortly after the memo was released, Attorney General Eric Holder said that this "important shift [would] ensure that the protections of the Civil Rights Act of 1964 are extended to those who suffer discrimination based on gender identity, including transgender status."

Almost overnight, the notion that anyone—not just those who had been diagnosed with medical conditions—could change his or her gender at will entered the mainstream. Once again, anyone who objected was deemed a bigot.

Just one year later, the Supreme Court handed down a flowery extralegal decision in the case of *Obergefell v. Hodges*, which concerned the question of whether state laws restricting marriage to one man and one woman were constitutional. Today, the text of that decision stands as a monument to the new age of enlightenment in which Obama and his administration believed we were living.

Writing for the majority, Justice Anthony Kennedy declared that any law denying the right of gay couples to marry violated the due process and equal protection clauses of the Fourteenth Amendment. As Justice Antonin Scalia, who dissented along with three other justices from Justice Kennedy's opinion, wrote, the words of that opinion lacked "even a thin veneer of law."

What was so shocking about *Obergefell* is that Justice Kennedy and the liberal justices had the confidence to baldly assert *as law* what were instead highly political progressive beliefs and assumptions. It's worth a read to see how deep this confidence ran. Justice Kennedy began by just stating, without argument, that "the nature of marriage" extends to "all persons, whatever their sexual orientation," and "there is dignity in the bond between two men or two women who seek to marry." This so-called nature of marriage was of course not natural in the sense that mankind for thousands of years understood it. According to Justice Kennedy, at its most basic level, "marriage responds to the universal fear that a lonely person might call out only to find no one there." With this fundamental redefinition of marriage as the desires of two "persons" for companionship, Justice Kennedy applied it to the case, announcing changes in the law as if they were edicts pronounced over all of history. "The limitation of marriage to opposite-sex couples may long have seemed natural and just, but its inconsistency with the central meaning of the fundamental right to marry is now manifest." Religion, tradition, and common sense had to be cast aside by Justice Kennedy's "better informed understanding" that had taken hold among the liberal elite.

But the fact is that the Constitution does not say anything about dignity, certainly not in the sense that it was used in *Obergefell*.

Since the founding of our nation, the people of the states have always had the right to determine their own laws, so long as those laws do not violate the Constitution. Clearly, the laws at issue in *Obergefell* did not violate the Constitution. If the activists who petitioned for the case to be taken all the way up to the Supreme Court wanted to change laws outlawing same-sex marriage, they should have persuaded their fellow citizens, as is customary in a democratic republic such as ours.

But they didn't.

Instead, they relied upon five unelected judges to create a right that did not exist before they made their ruling. This is not how judicial interpretation is supposed to be done. But in the new era of enlightenment that the liberal elite of this country believed we were living in, this notion of progress at all costs became the norm. Rather than allowing the people to vote on issues, or of allowing the states to govern themselves as they are supposed to be able to, the federal government expanded its power in an attempt to bring about a revolution. They were sure, as all revolutionaries have been since time immemorial, that their actions were necessary to safeguard the well-being of the citizens of this country. They believed they were doing the right thing.

In so doing, they created a new era in the United States—one in which we are still very much living today.

Cultural Marxism

All of this has led to a new divide in American politics. Rather than between liberals and conservatives, the divide seems to have emerged between ordinary people who live in the real world—the ones who have jobs, interact with other people on a regular basis, and want to raise their families according to traditional values—and people who live in a fantasy world. The political divide today is between common sense and crazy.

Politics used to be about differences in important but ultimately prudential judgments. Republicans wanted lower taxes, Democrats wanted higher taxes. Republicans wanted more defense spending, Democrats wanted less defense spending. Reasonable people can disagree but compromise, and even have a productive politics based on these differences. When the liberal elite began seeing the rest of America as its enemy, however, that changed.

The ones who live in the fantasy world are often the same people who were taught by crazy Marxist professors in the late 1990s and early 2000s. But now they have grown up and begun working at major corporations, writing editorials for major newspapers, and engaging in record amounts of online activism. Most of this "activism," of course involves little more than getting outraged at the latest "intolerant bigot" on Twitter, whom they will promptly try to cancel to satisfy their fellow online activists.

This divide does not play out simply on economics or politics. Rather, it is a divide between people who love this country and want it to prosper and those who see the United States of America as a fundamentally racist, sexist nation that must be radically changed or dismantled if it wants to survive. These people call for the nation not only to reject our traditions and our values and the time-tested principles that over the course of 240 years have made this the most successful republic in human history but also to reject the lessons and fundamental truths of what works and what doesn't from 5,500 years of recorded human history.

Every generation believes they have it figured out. Every generation believes that the one that came before them had it wrong—that everyone before you was ridiculous, but that somehow *you* are right, and that you've figured things out after thirty-five or forty years of life that have eluded everyone else who's lived before you.

This is where we get misguided notions such as "equity," a buzzword that has replaced equality as the standard we should strive for as a society. According to the laws of equity, all people should be viewed as a collection of handicaps and grievances, and our leaders should structure society so that those grievances can be addressed. The goal here, as many commentators have pointed out in recent years, is not equality of opportunity—which is what all societies should strive for—but equality of outcome. As the writer Ibram X. Kendi has written in his bestselling book *How to Be an Antiracist*, "When I see racial disparities, I see racism." In other words, any difference in status or wealth between two groups can be explained via racism.

This might sound insane, but it is nothing compared to the solutions that these same people propose. According to this new class of American Marxists, we must pass policies that do not treat all people as equal but make race a primary factor in our policymaking. In the words of Kendi, "the only remedy to racist discrimination is antiracist discrimination."[25]

The goal here is not equality. It is a perverse outcome whereby everyone has a different amount of status and wealth according to how much people who look like them may have suffered in the past. This is Marxism with a particular twist—one that could only have come in the twenty-first century in the United States.

Occasionally, when I speak about these people as American Marxists, I'll get skeptical looks. Some people have suggested that this word is hyperbole—something I use to scare people into action. But it's not. Marxism is more than just socialism. Socialism is an economic model; Marxism, however, is a *power* model.

Socialism, at its core, is about government control of the economy. Marxism, on the other hand, is about controlling everything. It is a power structure. Marxists believe at a fundamental level that people need to be controlled, and that they are the ones who should do it. The people, in their view, need to be controlled, and they need to be controlled by a small handful of people who are better, smarter, or more moral. Because if you don't control people, you'll get unequal outcomes. People will hoard wealth, discriminate against one another, and they won't allocate resources in the proper way.

Marxists believe we have to have control over everything. They believe we have to have schools that teach the values of Marxism from a young age. Education is important to them, but indoctrination is more important. It is important for them to get in front of people when they're very young and begin to mold and shape their lives. They cannot allow parents to interfere in this process. In fact, they teach children that their parents probably don't know what they're talking about, that their parents are a product of an old era. Parents, according to the Marxists, don't know the truth; only *they* know the truth.

The Marxists believe that even the schools aren't enough. They have to get the whole society to support them in their endeavor to bring about radical change quickly. So they need to make sure that the entertainers, the cultural influencers, and everybody else also reinforce these messages. Members of older generations don't know what they're talking about; all these old-fashioned things are not only no longer useful but evil. They're bad for young people, and therefore young people should cast them aside.

Marxism, of course, has never tolerated countermessaging. It has never been particularly good at accommodating the kind of dissent or argument that has made the United States of America the greatest nation in the history of the world. Anyone who spoke out against the regime in Cuba, for instance, was labeled a "subversive." In some places, they were called "counterrevolutionaries." Today in the United States, anyone who speaks out against the prevailing liberal orthodoxy—the one that began to be foisted upon the people of this country as soon as President Obama was elected a second time—are called hateful, dangerous, or "extremists."

As of the summer of 2022, the president of the United States demonstrated that he was comfortable calling these people "neo-fascists." It's worth noting, of course, that during President Biden's address in Philadelphia he did not just use this term to refer to the people who stormed the Capitol on January 6, 2021. He used it to describe anyone who did not support "a woman's right to choose" or "same-sex marriage." According to that logic, he and Barack Obama were semi-fascist less than a decade ago; so were a majority of the people who consider themselves authorities on what constitutes enlightened progressivism today.

This is what happens when one side of the political spectrum begins to believe that they are engaged in a radical project of remaking the country they are supposed to serve. When radical change comes too quickly to a society—and, more importantly, when it comes from a small group of deranged radical activists rather than from the people—fissures begin to develop among the people.

In a sense, Marxism has always existed in the United States. Throughout the twentieth century, there have been professors and groups of activists who operate on a Marxist framework. But for the most part, this framework has always been tied to socialism, which is an economic model that prioritizes conflicts between people from different income levels. The shift in our culture came, it seems, when these same Marxist activists substituted economic differences for racial ones.

This hyperawareness of racial differences certainly began during the second term of President Obama; but more than any other phenomenon, it has metastasized in the years since to become one of the most dangerous threats we face today.

Chapter 9

AMERICAN MARXISM IN ACTION

Before the summer of 2014, few Americans had heard of Ferguson, Missouri. The city was small and quiet, a community of about twenty thousand people that sat just outside St. Louis.

That changed on August 14, when a police officer pulled up beside two young black men who were walking in the middle of the street and asked them to use the sidewalk. According to accounts that would be given later, the two young men refused and exchanged words with the officer. At some point one of the boys—Michael Brown, whose name would soon become famous—attempted to grab the officer's service pistol, which, after a short struggle and pursuit, the officer fired, killing the young man.

This was a tragedy. Anytime there is a tragedy, someone, somewhere will attempt to exploit it for their personal gain. Sometimes the gain is merely financial, but other times it is to gain power—political, industrial, or even criminal. The sad truth is that Brown's death was ripe for exploitation because America's history with race is complicated, to say the least.

But the best way to understand this history is through the words of Dr. Martin Luther King Jr.: "When the architects of our republic wrote the magnificent words of the Constitution and the declaration of Independence, they were signing a promissory note to which every American was to fall heir," he explained while standing on the steps of the Lincoln Memorial. "This note was a promise that all men would be guaranteed the unalienable rights of life, liberty, and the pursuit of happiness." In fifty-four words, King explained the inherent goodness of America's founding ideals.

We remain in a continued pursuit of that "more perfect Union." And we cannot be afraid of that conversation. As I told a crowd in New Hampshire in 2015, "We do need to address the reality in this country that there are millions of people who feel that because of the color of their skin, they're followed at the mall, they're treated differently. If a significant percentage of the American family feels this way, it's an issue. We have to talk about it."

But as we strive forward toward Dr. King's nation where people "will not be judged by the color of their skin but by the content of their character," there are forces looking to drag us backward to a time when the color of our skin was our identity, the sound of our last name dictated our political allegiance, and we saw ourselves not as Americans but an ethnic group simply living in a place called America.

When Michael Brown died, we fell into an all-too-familiar pattern—candlelight vigils that turn violent as the sun sets, followed by calls for calm. The cycle repeats until the national media realizes ratings are slipping and energy on the ground fades away.

But the uprising in Ferguson, Missouri, was different in several key ways that were hard to pinpoint, at least initially. First, the outrage and destruction seemed to ramp up much faster than in the past. In the years since Barack Obama first took office, a notion had taken hold that the United States was a dangerous place for minorities—that people of color, especially black people, were somehow in danger of being killed by police every time they walked out the door.

This was an idea created and pushed by social media, which was just becoming popular at the time the Ferguson riots began. For the first time, anyone with a Twitter account could shoot videos of police interactions that went wrong—or, stripped of all context, appeared to go wrong—then show those videos to millions of people around the world. This created the illusion that bad police encounters happened every few seconds. It also allowed online activists to editorialize about these incidents before full investigations could be conducted. Sometimes the noise created on Twitter became so loud that legacy media institutions reported on it as well, creating a fact-free media firestorm where one did not exist.

Second, we saw the rise of new national groups, who began to drive the debate and, in some ways, bypass the local leaders on the ground. Black Lives Matter is perhaps the most prominent example. The relatively new group exploded onto the scene—and, as we would see years later, exploited these situations for personal financial gain as well—accelerating a national-level debate. It quickly became an umbrella group of sorts allowing people to claim crazy things, advance radical ideas, and do it all without any skepticism from the legacy media that was enjoying soaring ratings.

Expressing outrage is easier than solving problems, of course. What many refused to talk about, even to this day, is how the same de-industrialization that savaged the Midwest also destroyed some of America's greatest cities. When those good jobs moved overseas, far too many families were left behind in cities that would soon begin to decay. For decades, Democrat mayors and city council members allowed the problems to fester and grow. Drugs. Homelessness. Violence. It is no wonder new businesses did not move into these once great cities. And in the absence of legitimate economic activity, illegitimate activity flourished. As things grew worse, Republicans and Democrats responded with a mix of welfare and lectures about learning to code.

Brewing just beneath the surface was an alternate explanation for the ills of America, particularly those of the black community. It couldn't be a matter of trade policy or simple neglect; the culprit, we were told, was now systemic racism. It would not take long for this race-obsessed, neo-Marxist narrative to become mainstream.

During the same period, books and articles began to emerge that signaled a willingness on the part of our major institutions to adopt what would soon become known as "critical race theory." In time, we would all come to know the inner workings of this strange ideology all too well. We would find out that critical race theorists believe that the world is nothing more than systems of interlocking oppressions, and that race is the paramount factor in determining a person's destiny.

Whereas men such as Dr. Martin Luther King had dreamed of a world where race did *not* matter, and where all people would be able to live in harmony regardless of their skin color, the people who believed themselves to be his heirs began fighting for exactly the opposite.

In 2014 the writer Ta-Nehisi Coates published an article in the *Atlantic* titled "The Case for Reparations," which argued that black Americans should be paid somehow for the sin of slavery. A few years later, he wrote a bestselling book titled *Between the World and Me,* which dealt explicitly with the shootings of black men by police officers. In that book, he said that the circumstances of these shootings didn't matter, and that police officers who kill black men are "merely men enforcing the whims of our country." Later in the same book, he described "whiteness" as "an existential danger to the country and the world."

In the years to come, this view of the United States would become extremely popular on the mainstream left. For his work on racism and whiteness, Ta-Nehisi Coates was given a MacArthur Genius Grant, which came with a large cash prize and the status of being called a literal "genius" by the establishment. He would interview President Obama several times, always speaking to him about the legacy of race and racism in this country.

If this strange obsession with race were confined to our borders during those years, it would have been one thing. But it was not. In fact, it seemed that every time President Obama went overseas to give a speech, he managed to work in the sins that our ancestors had committed in the United States.

During those speeches, the world was listening. That included not only Germany, the United Kingdom, and other allies, but China and Russia as well—two countries that had been looking for weaknesses in the United States for years already. With this elite-generated obsession about race, they found it. In fact, Russia actively exploited the tensions during the 2016 presidential election. They didn't need to steal any state secrets to do so; they just needed to turn on the television and listen to the outgoing president.

Sadly, these two countries would have recognized better than anyone else the race-obsessed Marxist ideology that had taken over the American left. Rather than arguing that economic classes must be constantly at war with one another, these new American Marxists insisted that history was nothing more than an eternal battle between the races—one that so-called white supremacists had been winning for too long. They spoke in similar language of a revolution that needed to happen—which they often called a "reckoning"—and proselytized about it just like good Marxists.

During the George Floyd protests that shook this nation to its core during the summer of 2020, we would see this ideology on full display in the streets of our cities. We would also see it in the offices of our major newspapers, corporations, and classrooms. It was during these protests, which soon became riots, that an explicitly Marxist organization known as Black Lives Matter would rise to prominence in American life, taking in donations from major corporations and private citizens all over the country.

Just a few decades earlier, when communism had just collapsed in Europe and the world was well aware of the dangers posed by radical Marxist ideology, this would have been impossible. But in the years since, as we moved on and forgot about the danger inherent in revolutionary thinking of this kind, our awareness of that danger went away.

The story of how quickly—and how easily—this happened should serve as a warning to everyone.

Theory and Practice

In the years since Karl Marx first outlined his theories of economics and political power, several nations in the world have attempted to implement them. With almost no exceptions, each one of these nations has found out that Marxism, like socialism, leads inevitably to ruin.

By now, the reasons for this are clear. The entire concept is built around seizing power. Of course they don't say it that way, but as a practical matter that is what must happen. These systems require uniformity and conformity. If someone doesn't buy into the system, the system cannot work. The only way to ensure compliance is a strong government and a weak people. And as we see today, even in a nation as open as America, the government can still cajole social media companies to silence speech and shape debate. That behavior prompts a backlash and, when paired with idiotic economic policies, becomes unsustainable. That is a lesson people across the globe, including many new Americans from failed states like Venezuela, have learned, but it is one too many Americans choose to ignore.

Of course, American Marxism looks different from Marxism in China and Cuba. It is given a patina of authenticity and legitimacy because certain institutions in the United States are willing to give comfort and aid to Marxist revolutionaries. This has been particularly true of our universities, the faculty of which usually dwell in the realm of theory and grand historical ideas rather than the practical applications of those ideas. In fact, it was a group of professors at Harvard Law School who, in the mid-1970s, began playing with the ideas of Karl Marx in the articles they were writing for various legal journals around the country. It started small. Professors would mention systems of oppression and the eternal struggle between the haves and the have-nots in their lectures. They would write long law review articles that cited Marx, Engels, and other radical left-wing thinkers.

But soon things began to spiral in a strange—and, sadly, recognizable—direction. By the end of the 1970s, these scholars were writing articles full of citations to theorists such as Jacques Derrida, who believed that language was effectively meaningless, along with every structure in society. They were claiming that the law, in the words of the scholar Derrick Bell, was nothing more than a tool that powerful—read: white—people use to oppress everyone else, especially people of color. In time, this approach to the law came to be known as critical legal studies, and it spread to several major universities across the country.

As this ideology migrated, it picked up terms from other disciplines along the way. When it touched humanities departments, for instance, it began to take on the explicit language of fields such as black studies, which were hyperaware of race and racism. Thus, the discipline known as critical race theory was born. In a book-length introduction to the concept—which, if you believe today's prominent liberals, existed (and was, in fact, a very good thing) right up until the moment that conservatives began expressing their concerns about it—critical race theory "questions the very foundations of the liberal order, including equality theory, legal reasoning, Enlightenment rationalism, and neutral principles of constitutional law." It is different, according to the author, from "traditional civil rights," which embrace "incrementalism and step-by-step progress."

In the view of these scholars, "incrementalism and step-by-step-progress," the processes through which nearly every major victory against intolerance in this country has been achieved, were not good things. Rather, they were impediments to a glorious political revolution—one they were sure, as *all* dedicated Marxists before them had been sure, was right around the corner. Their goal was not to work within the American system to improve the country; it was, rather, to destroy the country and then build it back up according to their own radical beliefs. As anyone who's studied history knows, Marxist revolutionaries are very good at destroying a country. But they have never quite been able to get the hang of building it back up, which involves compromise, reasoned debate, and accommodating the beliefs of those who don't completely agree with you.

Throughout the 1980s, the tenets of critical race theory continued to spread throughout our nation's major universities. Thousands of lawyers, teachers, and even doctors graduate every year after having been exposed to courses that had this strange, half-baked philosophy worked in somehow. And given that this soft academic version of Marxism did not have to work in the real world, it grew even stranger and further removed from reality as it moved through various academic departments. Suddenly, there were Marxist ways to read the great books of our past, analyze unemployment statistics, and write term papers. There was nothing that this philosophy could not touch, because anything could be twisted to fit the grand narrative of eternal struggle upon which all Marxist thought rests.

Theoretically, the collapse of the Soviet Union in the late 1980s should have put the final nail in the coffin of Karl Marx's ideology. At a time when Communist governments were crumbling all over the world, you would think that the few remaining proponents of Marxism—especially the ones who were living in the United States at the time, enjoying the fruits of the capitalist system that Marx spent his entire career railing against—might have renounced their faith, as it were, and moved on.

But that's not what happened.

Instead, the few professors and ideologues who still subscribed to Marxist principles hid themselves even further away from society and continued expounding on the man's crackpot theories. In October 1989 a reporter from the *New York Times* covered this development in some detail under the headline "The Mainstreaming of Marxism in U.S. Colleges," which appeared in the paper's education section. At the time, this article must have seemed like a whimsical look at a strange development on our campuses. That it appeared on page B6 above a longer story on seniors attending college for the first time highlights how minor the editors of the *Times* believed this new trend to be.

Today, it seems downright chilling.

"As Karl Marx's ideological heirs in Communist nations struggle to transform his political legacy," the article says, "his intellectual heirs on American campuses have virtually completed their own transformation from brash, beleaguered outsiders to assimilated academic insiders."

Over time, as the world began to forget the horrors of true Marxism, a watered-down, ever-shifting version of it grew more and more popular on college campuses. Every year hundreds of thousands of undergraduate students signed up for courses that not only took the philosophy of Karl Marx seriously but also listed his writings at the tops of their syllabi as required reading. Before long it became nearly impossible to take courses in literature, economics, or political science without having a small amount of Marxism thrown in for good measure.

It is no surprise that during this period, instances of campus activism rose sharply. The causes for which these protests were happening also grew stranger. In the 1990s, when I was a student in law school, I remember activist groups at my own university protesting about divestment from fossil fuels and nuclear weapons. I didn't agree with these causes, but it was clear that the students raising their protest signs believed they were helping the world become a better place.

Less than two decades later, I saw students expressing similar outrage over Halloween costumes and professors who had used words they didn't like in classrooms. I read stories about students demanding to have "safe spaces," where only people of certain races would be allowed to go. During this period, it became increasingly common for students to ask for "trigger warnings" to be placed in front of classic works of literature because those works contained violence, sexism, and racism. Any professors who resisted ran the risk of being reported and fired for their lack of dedication to the cause of equity and antiracism.

The echoes of the Marxist society that my neighbors fled and that had ruined my parents' homeland were impossible to ignore.

The only difference was that now, rather than focusing obsessively on the struggle between owners and the working classes, these students were focused on the eternal struggle between people of different identity groups. In truth, they really cannot focus on economics because many of the leading intellectuals (if you can call them that) and activists are very well-off financially. Some of the primary texts of this era were written by the left-wing professors who had gone underground after the defeat of Marxism in Europe. Two of the most famous are Kimberlé Crenshaw and Derrick Bell, two scholars with Marxist roots who were instrumental in developing critical race theory.

According to Kimberlé Crenshaw in particular, society should focus not only on racism, sexism, and other types of discrimination but also on "intersectionality," a term that refers to the various ways that people can be oppressed. According to Crenshaw's work, people should consider all the oppressed identities that might apply to them, then complain endlessly about the ways that those various identities "intersect."

There are various names for this sort of thinking. Some writers have called it "grievance studies." Others have called it "wokeness," adapting a term that first became popular on Twitter and turning it into a pejorative. Whatever terminology you choose, it is clear that this political ideology has very little in common with a nation founded on the freedom of speech and the fundamental truth that all men are created equal. These new "woke" activists seek to censor speech and to dwell constantly on the politics of racial differences.

Recently, the essayist Wesley Yang has coined the term "successor ideology" to describe this loose, often contradictory set of illiberal ideas that seems to have taken over the college-educated radical left. This ideology, he writes,

> *posits what I call a "unity of all oppression" narrative and sets itself in opposition to what it calls "Eurocentric cisheteronormative patriarchy," which is a pretentious way of saying "the rule of straight*

white men." The ideology says that whites are privileged over non-whites, that men are privileged over women, that the able-bodied are privileged over the non-able-bodied, that heterosexuals are privileged over homosexuals, and that every one of us has a unique experience of both privilege and oppression structured by the dimensions along which we are privileged or oppressed that in sum accounts for who we are and where we end up in the world.

According to this account of social reality, one of the goals of legitimate institutions, and part of the basis of their mandate to rule in the age of ideological succession, is that they must work to dismantle those implicit hierarchies in all their myriad guises.

For years we heard that students at elite colleges would stop their over-the-top activism once they graduated and had to join the workforce. Anyone who published articles about the increasing trend toward illiberalism among students was told that they were overreacting, or looking for problems where none existed. But they all ignored the fact that Marxism had become extremely entrenched on these campuses—and that part of the curriculum involved changing institutions to better align with neo-Marxism, not the other way around.

When these students entered the workforce, that's exactly what happened.

Changing Times

During the fall of 2019, hundreds of people lined up on the sidewalk for a print edition of the latest *New York Times Magazine*. The cover story in that edition, which had already been published in full online, was a long piece introducing something called the 1619 Project. According to Nikole Hannah-Jones, the writer who had conceived the project from within the *Times*'s newsroom, this collection of essays from journalists, writers, and a few academics was meant to put forward "a new origin" story for the United States—one that centered on the date that slaves first arrived in North America rather than the date of America's founding.

The praise from other liberal media outlets was immediate and intense. That year, Hannah-Jones was awarded a Pulitzer Prize for her work on the piece, an award known around the world as being the highest honor a journalist can receive. But unlike most of the work that had won the prize in the past—the articles written by Bob Woodward and Carl Bernstein that took down President Nixon, for instance, or Neil Sheehan's exposé about the Vietnam War, the Pentagon Papers—the work that was published in the *New York Times Magazine* in August 2019 did not concern itself too deeply with the facts.

Shortly after the piece was published, several professors of history from around the country noted egregious errors in it. In a public letter, they noted that several of Jones's central claims—including that the American Revolution was fought primarily to preserve the institution of slavery—were wrong, and that her writing "displayed a displacement of historical understanding by ideology." A few months later, a researcher who had worked on the piece said that although she had pointed out several errors, the editors at the magazine had ignored them and gone ahead to publication anyway.

Anyone who's studied the history of radical movements will see the pattern here. Although the reporting was not strictly true, or particularly good, it was backed by the regime because it was good for the cause. For the next few years, the editor of the *New York Times Magazine* would spend a great deal of his time fending off accusations from serious scholars that the ideas contained within the 1619 Project were not based on any historical data whatsoever.

It is annoying that the 1619 Project gets its history wrong. But it is downright dangerous how it gets its understanding of America wrong. It is an undeniable truth that the slave trade and slavery are important parts of American history. Nobody in America denies this or fails to understand it. But slavery is not the central fact of the

American experiment. The DNA of America comes, as generations of Americans have understood and taught their children, from our founding principles: that all men are created equal and endowed by their creator with certain unalienable rights. It is because these ideas are the DNA of our country that every positive social movement in American history has been led by Americans calling on our nation to be true to its founding principles. This is how slavery was ended. This is how the vote was given to women. This was the objective of the civil rights movement.

Progress in our country did not come because a foreign power invaded us and imposed on us an end to slavery or equal rights for women. Progress came when Americans—often deeply religious Americans—pointed out how we were failing to live up to the ideals of America. This is what we should teach our children. Instead, the 1619 Project tries to teach them that America is rotten down to its core—that America was conceived in slavery. What a demoralizing message to send. If Martin Luther King Jr. had believed that, he wouldn't have called on America to do better. He would have demanded America be torn down. And that's precisely what radical progressive leaders say today.

If this kind of thing were restricted to a magazine connected to a major newspaper—one that most people don't read anyway, given that it comes once a week with the print edition of the newspaper—that would be one thing. But as the months went on, it became clear that virtually every liberal media institution in the country had become infected with the same thinking. Rather than growing up and changing their thinking when they got real jobs, the campus protestors who had made headlines during the presidency of Barack Obama—the ones we were assured would grow out of their ridiculous neo-Marxist ways of thinking—began to change the institutions they worked for. In a matter of years, the liberal bias at institutions such as the *New York Times* grew into a shocking propensity for Marxist thinking and left-wing activism.

This was reflected in the tone and content of stories, as anyone who reads the paper regularly can tell. But it was also reflected in the internal politics of the newsroom. In 2018, for instance, after watching an Asian American figure skater win a gold medal in the Olympics, the opinion editor Bari Weiss tweeted this message: "Immigrants: we get the job done." The line, a reference to the musical *Hamilton,* seemed inoffensive to most normal people. But the staff of the *New York Times,* like the staff of most major media institutions at the time, was anything but normal. Soon furious messages were being sent back and forth on the newsroom's Slack channel, some of which spilled over into the press. For her crimes, Weiss was accused of white privilege and outright racism.

A tendency toward this kind of zealotry was already present before the murder of George Floyd in May 2020. Once that incident happened, however, the nation learned just how entrenched movements such as critical race theory had become in our institutions—especially in the mainstream media. For weeks, as cities burned and protests devolved into riots, networks such as CNN covered the carnage with a positive spin. They seemed determined to make the protestors and rioters look good at all costs, and to make anyone who suggested we protect civilians, property, and police officers seem racist. I'm sure most people remember the sight of a CNN correspondent standing in front of a street where cars had been toppled, windows had been broken, and fires were still burning, with the chyron reading: "Fiery but mostly peaceful protests after police shooting."

Again, the movement was more important than facts. Anyone who disagreed was quickly branded an enemy of the revolution. On June 3, for instance, my colleague Senator Tom Cotton published an op-ed in the *New York Times* suggesting that President Trump could, if he so chose, send in the US military to stop the looting, rioting, and violence that was unfolding in the streets—not just in New York City but all over the country.

In response to the article, the writers and reporters of the *New York Times* went crazy. So did the rest of the New York media world.

The explosion began on the internet, but it soon spread out into the real world. Within a few hours, this message began showing up on the Twitter feed of just about every reporter at the *New York Times*, right above a link to Senator Cotton's article:

Running this puts black @NYTimes staff in danger.

None of them, to my knowledge, explained exactly *how* running the piece had put black staff members of the *Times* in danger. Hyperbole and outlandish claims are increasingly common in today's political discourse, but what stood out most was the hypocrisy.

The same people who expressed horror at protecting our cities in 2020 were quick to celebrate the deployment of the National Guard in Washington, DC, in 2021. Military-style checkpoints, razor-wire-topped fencing, and massive law enforcement raids across the country were deemed essential to saving American democracy. For a while, America's capital looked more like a fortified compound than a shining city on a hill. And America's Marxist left was totally fine with it. Remember what I said about control?

Of course, these are also the same people who celebrated and honored Fidel Castro, a man who routinely used force to punish and silence all those who dared to speak out against his Marxist regime. They also wear T-shirts featuring Che Guevara, a racist, homophobic mass murderer.

What this boils down to is an intense anti-Americanism. That—and that alone—is the guiding principle of this new American Marxism: a complete and utter destruction of our country's ideals, values, and institutions.

Just consider the fallout at the *Times* over Cotton's op-ed. By the end of the following week, editor James Bennet had resigned. The paper also decided to add a lengthy note to the beginning of the piece, warning sensitive readers about what was inside. Now, compare that to the complete lack of reaction, response, or mea culpa to other *Times* pieces, like the one called "Pedophilia: A Disorder, Not a Crime" from October 2014. You won't find any notes about how it might put children in danger. As of this writing, you're also able to read an op-ed from Sirajuddin Haqqani, the leader of the Taliban, called "What We, the Taliban, Want," without having to sift through a note about how the piece "falls short of the *Times's* editorial standards."[26] That one, by the way, ran in February 2020, just a few months before Senator Cotton's.

Apparently the trend of censorship, intimidation, and silencing didn't end with Senator Cotton. On July 15 Bari Weiss—one of the few writers at the paper who had seen the left's assault on free speech coming—announced that she was resigning. In a letter posted to her website, she detailed months of abuse and bullying by supposedly liberal colleagues at the paper, and included this paragraph, which is chilling:

> *Twitter is not on the masthead of The New York Times. But Twitter has become its ultimate editor. As the ethics and mores of that platform have become those of the paper, the paper itself has increasingly become a kind of performance space. Stories are chosen and told in a way to satisfy the narrowest of audiences, rather than to allow a curious public to read about the world and then draw their own conclusions. I was always taught that journalists were charged with writing the first rough draft of history. Now, history itself is one more ephemeral thing molded to fit the needs of a predetermined narrative.*[27]

When this letter was published, the most radical elements of the Democrat Party had not yet won the White House. Many believed they would never be able to do so, given the support they were showing for open rioting and looting in the streets of our cities. It did not seem right that a party whose members seemed to hate everything that the United States of America stands for—who, indeed, make some halfhearted attempt every few years to completely rewrite the history of this country from its founding, refocusing the whole thing according to some strange neo-Marxist ideology—should ever be in charge of *running* the country.

But in November 2020, largely thanks to a historic effort by Big Tech companies, media institutions, and the intelligence community, Joe Biden managed to win the White House. Many Americans probably thought they were voting for normalcy. It had been a difficult several years between the Covid pandemic, the constant outrage over everything Donald Trump did, and President Trump's own penchant to stir the pot by saying outrageous, and sometimes offensive, things. But we wouldn't get normalcy. Instead we got the most radical, Marxist presidency our country has ever seen, and the radical agenda went completely mainstream.

Chapter 10

A TIME FOR CHOOSING

Name your favorite Republican. It has become a predictable and pretty meaningless debate question over the years because almost every Republican says Ronald Reagan. There is a really good reason for that, of course. After all, Reagan led the United States to victory in the Cold War. A mix of inspiration and determinism not only led to the downfall of the Soviet Union but stopped the spread of Marxism across the globe.

But there is a part of Reagan's story we conveniently overlook today. He was considered to be a radical, far outside the mainstream of the day's political establishment. President Jimmy Carter warned the outcome of the election would determine "whether we have peace or war." The *New York Times* piled on in 1981, writing, "The man, whom we call a conservative, turns out to be downright radical."

We know how the story ends. Reagan overcame an establishment class that hated him, fundamentally reshaping not only the Republican Party but also the country's entire political framework for a generation. In 1984 he won a record forty-nine states and 58.8 percent of the vote. For the next three decades, millions of Americans referred to themselves as "Reagan Democrats."

In many ways, this was the beginning of political engagement for me. My family and I had just moved to Las Vegas in 1980, and football was the only thing I found more interesting than politics. Senator Ted Kennedy's challenge to incumbent President Jimmy Carter was a galvanizing moment. I was a Kennedy fan—living is learning, and thankfully I've been able to do a lot of living—and his concession speech was inspiring.

Even to this day, there is a line that jumps out to me: "Our cause has been, since the days of Thomas Jefferson, the cause of the common man and the common woman."

In that simple line, Kennedy reaffirmed the inherent goodness of our nation's founding and our task, as leaders, to pursue the common good. It goes without saying he'd be run out of today's Democrat Party for praising a founder and excluding people who don't identify as either a man or a woman. But he spoke to the fundamental truths of our nation.

By the time I was in fifth grade, I wrote a paper praising Reagan's effort to rebuild our military and restore our national strength. My grandfather—a man who had seen his country succumb to the tyranny of Marxism—played a huge role in that evolution. But perhaps more importantly, he urged me to study history, learn from it, and then use that knowledge for good.

And here is where many of the politicians invoking Reagan's leadership and legacy have it all wrong today: Ronald Reagan would never have wanted us to simply repeat his playbook from the 1980s.

Reagan understood better than anyone that times change. In fact, his presidency is a testament to changing thoughts, beliefs, and policies.

"I never meant to go into politics," he said during his farewell address to the nation in 1989. "But I was raised to believe you had to pay your way for the blessings bestowed on you. I was happy with my career in the entertainment world, but I ultimately went into politics because I wanted to protect something precious."

Look, I understand the temptation. Every Republican politician for the last three decades has succumbed to the siren song of "What Would Reagan Do?"—myself included at times. How could we not? He was one of the greatest leaders and presidents in American history.

The temptation to channel Reagan gets even stronger when we

consider the stunning parallels between Jimmy Carter and Joe Biden, another president who has brought us record inflation, weakness abroad, and division at home.

But when Reagan took office, the corporate tax rate was a stunning 46 percent. Today it is 21 percent. In 1980 the United States dominated global trading. Today, that title belongs to the Chinese Communist Party. Manufacturing went from providing around 20 percent of jobs in America to less than 10 percent. Out-of-wedlock births have more than doubled from below 20 percent to around 40 percent today, and at the same time, regular attendance of religious services has plummeted. The list goes on and on.

There are lessons to learn from Reagan—including many forgotten lessons, like his decision to impose a 100 percent tariff on certain Japanese electronics—but we are in a different time. We require new policies, new ideas, new institutions, and new leaders.

Conservative, Inc.

Looking back, it is shocking to realize how much of Washington, DC, was still stuck in the Reagan-era bubble when I arrived in 2011.

My first campaign for Senate was one of which Reagan would have been proud. Taking on a popular two-term governor of your own party was either radical or downright dumb, depending on your perspective. My opponent had all of the best consultants, all of the most coveted endorsements, and more money than he needed. All I had was an idea and a passion to serve. It was an outsider campaign that resonated with Floridians tired of hearing the same lines from establishment politicians over and over again.

In his "Time for Choosing" speech, Reagan quipped, "Our Democrat opponents seem unwilling to debate these issues. They want to make you and I believe that this is a contest between two men—that we're to choose just between two personalities."

He had it half right. The other half was that too many of our

Republican friends were also unwilling to look at hard policy questions and come up with interesting answers. Boring. Don't make waves. May the best hair or smile win.

Even though Reagan's smile was unmatched, he represented as much as anybody the notion that politics isn't just about optics but also about "communicating great things." He had a bigger vision for America and sought to tap into something brewing just beneath the surface. We'd see a similar dynamic thirty-six years after his 1980 presidential victory.

But I am struck now by just how well entrenched those former radicals had become in Washington by the time I arrived. Every policy meeting, every think tank, every lobbyist and interest group—all of them seemed to be a decade behind, part of what some referred to as Conservative, Inc.

Ronald Reagan had long since left office, but his legacy was being tarnished by those who claimed to be upholding it. Instead of bringing Reagan's optimism, humor, and conviction in America and her principles, Conservative, Inc. was on perpetual autopilot, applying little beyond Reagan nostalgia to every problem our nation was presented with.

As many immigrant families will tell you, Reagan's sense of boundless optimism was well deserved. America was and is a place like no other in human history. It truly is a shining city on the hill, a beacon of hope for the world, and a place where anything can happen.

It is also a nation in desperate need of a leader who understands that the threats to that vision are different from the threats Reagan confronted and overcame.

Having seen the American Dream become reality for my parents, I was eager to save it for my children and (someday) their children. I spent most of my first term exploring ways to do exactly that. I asked questions, pulled in experts, and pushed my staff to think outside the box.

I vividly recall one policy expert telling me not to worry because the American Dream was alive and well. Median household income had nearly doubled between 1990 and 2016, he said. The size of our economy had tripled. And the stock market was up 600 percent. It was morning in America. All that was left to do was to cut a few more taxes, pass a couple more free trade agreements, and slash some regulations.

Now, there is nothing inherently wrong with those policy ideas. Depending on how you do them, they can do a whole lot of good for a lot of people. But they missed the point. During an era of reduced taxes and increased free trade, American families suffered, dignified work disappeared, and communities crumbled.

America needed us to do more, so we did.

It began, as with most Republican efforts, with tax reform. Our tax code was dreadfully out of date, making America less competitive and investment in real things less desirable. That needed to change, but I was determined to make sure we did not leave working families behind. I told my party I would only support tax reform if we doubled the size of the Child Tax Credit for hardworking parents.

Washington and Wall Street were furious. The *Wall Street Journal* editorial board criticized me. Colleagues told me I was jeopardizing tax reform. Today, the precise details of this disagreement might be hard to imagine. They might even seem trivial, considering the issues we're dealing with in the present day.

It was embarrassing that all these people believed a 20 percent corporate tax rate was more important than letting parents keep more of their money. I told everyone that companies would do just fine with a 21 percent corporate tax rate, but families were not going to thrive if we kept taking more of their money.

Ultimately, after a weekslong high-stakes standoff with my own party, we succeeded in doubling the size of the Child Tax Credit.

That wasn't just success on paper. It mattered to real people almost immediately. Data from 2018 tax returns found that the number of people who received the credit doubled because we expanded eligibility. That meant more low-income parents were able to keep more of their tax dollars. And that data found that Americans in every single tax bracket benefited from the increased benefit, except those at the very top of the income scale.

It was an example of common-good capitalism at work—prioritizing our families over slightly higher corporate profits.

I was determined not to make it a onetime event.

Three years later we changed the law to prohibit Chinese companies from listing on American stock exchanges if those companies did not comply with our nation's transparency laws. Now, that may sound like common sense, but for decades Chinese companies were allowed to ignore US securities rules. The result was that investors, including retirees and pensioners, were often investing in companies that were lying about their financial records. Countless Americans lost huge sums of money as a result.

America now has access to audits from those companies—at least some of them—and those that do not comply will be removed from our exchanges. That will protect our retirees and pensioners from exploitation. It is the right thing to do. Short-term profit doesn't mean anything when we undermine the credibility of the market.

But there was another reason to remove these companies. When these companies receive investments from Americans, that money strengthens these Chinese companies. It provides them with capital to expand and conduct additional research and develop new technologies and weapons—yes, weapons that one day may be used to kill American service members.

There is absolutely no reason we should be funding that type of activity. The Chinese Communist Party wants to cripple America, and they have found a way to make us pay for it.

Just as Stalin predicted, "When we hang the capitalists, they will sell us the rope we use." This is worse. We are giving the Chinese the money they need to make the rope they are going to use to hang us. Cutting off the flow of resources to the Marxists in Beijing will take time, but we've taken the first step.

These are just two small examples of the beginning of a long-awaited policy shift within the Republican Party, one that was a couple decades overdue, but it was happening within the confines of a stale, broken system. The impacts would be meaningful, but the pace of change threatened to be glacial.

The Great Disruptor

Like many Americans, I was shocked when Donald Trump rode down the escalators in New York City to announce his presidential campaign.

At the time, I was two months into my own presidential campaign. I had launched it in April 2015 at the Freedom Tower in Miami, entering the race with a message of hope and optimism that seemed appropriate given the last two election cycles we had endured. According to the conventional wisdom in the Republican Party at the time, we needed another Reagan. The Obama economy was crushing hardworking Americans, and we'd become a laughingstock abroad. The pundits, consultants, and donors—all of whom did really well under Obama's economy—believed the path to Republican victory was boundless optimism in America's future. In other words, everyone needed to hear that they too could achieve the American Dream, and the only thing standing in their way was a Democrat in the White House.

Donald Trump's announcement took on a much different tone. America could still be great, he argued, but not so long as the corrupt, stale elites ran Washington. Looking back, I shouldn't have been as surprised as I was when so many Americans began to pack stadiums all around the country to hear longer, even more bombastic speeches from him. Here was a guy from the outside. He didn't have the best consultants. He didn't have the coveted endorsements. And while he had a hell of a lot of money, he certainly wasn't eager to spend it on a presidential campaign. What he did have was an intuitive understanding of how the American people felt and what they were looking for in a leader.

Watching the Trump campaign in action, I was reminded of my own first campaign for the US Senate in 2010. Of course, I didn't have billions of dollars or a private jet with my last name on the side. But I did have an outsider spirit that allowed me to connect with voters who felt that the government was not working for them.

If I am being honest, it was hard to keep that outsider spirit during my first term in the US Senate. Sure, I was still fighting for the American Dream—it was the reason I woke up every morning, suffered through the donor calls and fundraising dinners that are part of running a presidential campaign—but Washington has a way of rounding off the edges. There are consultants. Lobbyists. Fundraisers. Think-tank eggheads and pollsters.

Everything in Washington is designed to grind good ideas into the least common denominator. Sometimes that is required to pass legislation. But the art of negotiation and compromise is distinct from the art of inspiring a nation. Campaigns for president, and even the presidency itself, are about leadership, painting a vision for the future, and projecting confidence. They allow you to negotiate from a position of strength, backed by the American people and the confidence that what you're doing is good for the nation. All too often, Washington thinks "doing something" is the same thing as doing what is right. And voters were tired of it.

When Donald Trump won the presidency, we were on the brink as a nation. Despite the best efforts of our movement, the post-Reagan Republican Party allowed it to happen. In many ways, as we've discussed, they were actually complicit in the complete destruction of our nation's manufacturing base. They were complicit in the rise of Wall Street over Main Street. And they were complicit in turning a blind eye to the destruction of the family.

Two lines from Trump's announcement speech turned the Republican Party upside down and ushered in a generational shift in our nation's politics: "It is time to stop sending jobs overseas through bad foreign trade deals," and "It is time to close loopholes for Wall Street."

Over the next four years we began to take aim at Wall Street, multinational corporations sending jobs overseas, and Communist leaders in Beijing. The Republican Party became the party of the American worker, and the Democrat Party became the party of the American financier and TikTok influencer.

Donald Trump would be the first to tell you that he didn't accomplish everything he hoped for during his term. It was an era marked by Democrats' election denial, unfounded media inquisitions, and outright abuse of the impeachment process. We also had to fight against deeply embedded resistance within the federal bureaucracy and the Republican Party itself.

But there was no going back.

The party and the country had changed. Today, we are seeing a political realignment that is just as significant, if not more so, than the one Ronald Reagan confronted when he declared it was "time for choosing" in the late 1960s. Policy is evolving. New institutions are emerging. New coalitions are being built.

My reelection is proof that Republicans can build a multiracial working-class party if we reject the elite consensus and embrace the commonsense values that served our nation well for centuries. It is the blueprint to saving our nation, which is something I believe can still be done. I would not have written this book if I didn't, and I certainly wouldn't be serving my third term in the US Senate, working every day to ensure that the people who elected me would have a better life. But America is not going to be saved by those clinging to the pre-Trump Republican Party.

After I suspended my campaign for presidency, I heard from a lot of people who wished I hadn't done so. They were scared for the future of the country under Hillary Clinton. They were scared for the future of the Republican Party under Trump. But I reminded them of what I said in a speech at the conclusion of my campaign:

This nation needs a vibrant and growing conservative movement, and it needs a strong Republican Party to change the direction now of this country, or many of the things going wrong in America will become permanent and many of the things that make us a special country will be gone.

At the time I said those words, I believed we were on the verge of catastrophe. I didn't know whether President Trump would be able to prevail over Hillary Clinton. Very few people did. But handing the executive branch over to another Clinton wasn't an option. Thankfully, Americans felt the same way. Over the next four years, we got more done for hardworking American families than anyone thought possible, and arguably it was the most productive period since Reagan's first term. With four more years of that dynamic, we might have achieved my stated goal of a renewed, even stronger Republican Party to work for the American people.

Sadly, that is not what happened.

Biden's Jihad

Many Americans hoped two things would happen with the election of Joe Biden over Donald Trump. First, that the daily media-driven chaos that came to define the Trump presidency (tweet, hyperbolic reaction, impeachment, repeat) would finally come to an end. And second, the politics would fade a bit from our public consciousness, and we could just do the basic blocking and tackling for a while.

Neither were true.

Democrats and their stenographers in the media remain as obsessed with Trump as ever. He is good for their ratings. In fact, he may be single-handedly responsible for keeping the *New York Times* and MSNBC afloat over the years. Democrat operatives also needed Trump in the news to gin up political excitement. Trump helped them raise small-dollar donors and recruit activists.

But that came at a cost. It took more than "Trump bad" messaging to keep these radical Marxists engaged in politics. They were demanding results. And Joe Biden, the man who waited three decades to become president, was never going to tell them no. He needed their money, their activism, and their votes. In doing so, he unleashed some of the most radical and insane policies the country has ever seen.

During the lockdowns, Democrats and their media sympathizers yelled at me to "follow the science." That was code for "Shut up and don't question the government." That should be a warning flag for anyone with even a little common sense and a basic understanding of history. It set alarm bells off for me, and is one of the many reasons you see Hispanic Americans fleeing the Democrat Party. They hear the echoes of the authoritarian regimes they once fled.

But the point about science is broader. As soon as the lockdowns lifted—conveniently, once Democrats won the election in 2020 and were worried about blowback—they began a new crusade: trying to convince boys they could be girls and girls they could be boys.

Since Biden came into office, the cases of young children identifying as a different gender have exploded. As one example, the Montgomery County, Maryland, public school system reported a 582 percent increase in reports of "gender non-conforming" students in just two years! Nature is not changing, our culture is. Democrats and their radical allies are destroying truth, and in the process destroying the lives of thousands and thousands of young people. Democrats celebrate "gender-affirming surgeries" for young people, which means these people, often too young to even drink or rent a car, have cut off parts of their anatomy. Again, the results are irreversible. And every single day in schools across our nation, kids are being asked their pronouns, counseled and even encouraged to change their gender identity, and girls are finding boys in their bathrooms, locker rooms, and athletic competitions.

This is all considered "progress" in Biden's America, but normal people consider it evil, dangerous, and deeply damaging.

That alone would be enough reason to kick Biden and his ilk out of the White House in 2024, and for many people—including a growing number of Democrats—it will be the catalyst that has them voting for the eventual Republican nominee. But sadly, our problems are far greater.

That same woke spirit has thoroughly infected our nation's military. Rainbow bullets. Drag queen story hours. Stand-down orders to lecture our military about its inherent racism. The list goes on, and so too does the rest of the world. Because as the Biden people try to turn our military into a woke, politically correct agent of cultural change, the world is only getting more dangerous.

A disastrous pullout of Afghanistan led to a national embarrassment. It was heartbreaking to watch, and even more heartbreaking for our men and women who lost friends there. They saw this administration turn its back on everything they fought for. The weakness very likely emboldened Putin and led to the invasion of Ukraine.

If the world views America as weak, all bets are off. China may accelerate its timetable to invade Taiwan, which would give it control over vast sectors of the economy, including semiconductors. Iran may move even more aggressively in its efforts to kill foreign leaders it disagrees with. North Korea may increase its nuclear and ballistic missile tests, with one small mistake being a catalyst for war.

As scary as that situation is, most Americans cannot focus on it because their communities are overrun with crime and drugs. Joe Biden and Kamala Harris ran on a campaign of lecturing the police, siding with criminals, and downplaying the need for law and order. They supported letting hardened criminals out of jail early, released thousands convicted on federal drug charges, and allowed fentanyl to flow freely across our southern border.

They liked to paint a picture of everyone crossing the border as a refugee seeking asylum—an individual or family desperate for a new start. There are tens of millions, if not hundreds of millions, of people who would welcome a chance to live the American Dream. We cannot accommodate all of them. But hidden in the huddled masses are drug dealers, human traffickers, and cartel killers.

They flow freely into our nation, at the invitation of Joe Biden, Kamala Harris, and all these other lunatic Democrats who don't give a damn about America's future. Remember, they think America is a racist and irredeemable nation. It is not special and it doesn't merit preserving. In their eyes, a border is simply a tool of oppression.

And as they weaken us from within, they pay little attention to China. In fact, their policies signal to China that we will continue to be reliant on them for decades to come. Every single Democrat will complain about our reliance on oil from the Middle East—a problem we could drill our way out of—but they are willing to make us almost 100 percent reliant on China for solar panels and the batteries that power their precious electric vehicles.

Again, they don't actually care about America. Who cares if China is stronger than America? We're just a nation, nothing special in their eyes. Who cares if China is a Marxist atheist nation? That sounds good to them!

The 2022 midterm elections were a wake-up call—not for Marxist Democrats, but for Republicans. While a narrow majority in the House allows us to block Democrats from passing bad legislation, the failure to capture the Senate is an outright embarrassment and should cause us to do some serious soul-searching—the kind of soul-searching outlined in this book. We need Republican candidates who understand the threat to our nation and are willing to do what it takes to rebuild our great nation. This isn't a cult of personality or about owning the libs, it is about saving America for our children and their children.

A New American Century

I kicked off my 2016 campaign promising to bring about a new American century. Now, almost eight years later, I cannot help but think of Reagan's words back in 1967:

> *Freedom is a fragile thing and it's never more than one generation away from extinction. It is not ours by way of inheritance; it must be fought for and defended constantly by each generation, for it comes only once to a people.*

Today, it is not only our freedom at stake, but the very idea of America and the God-given truths it was founded on. Marxism has taken root in America and threatens to upend the greatest nation in the history of the earth. As we look ahead, our future—the world's future, really—will be defined by the great power competition between us and the Chinese Communist Party. There are only three broad outcomes, and only two are remotely acceptable.

To state what should be obvious: it is not acceptable for the twenty-first century to become the new Communist century. Unfortunately, that is also the most likely path at this moment. Too many in America's political, corporate, and cultural leadership remain scared of, beholden to, or infatuated with China. They have no interest in confronting the very real threat posed by the Chinese Communist Party, a regime that enslaves its people, erases cultures, hooks Americans on fentanyl, and steals everything of value it possibly can from other nations. We can already see the deaths, despair, and devastation in our communities because of China. And all of that happened before they were a near-peer competitor. Now think of the destruction if they beat us in advanced materials, artificial intelligences, and robotics. America's success—maybe even our survival—will rely on the grace of a Marxist regime that hates us.

Another possible path, though one I think is very unlikely, is the implosion of the Chinese Communist Party and the collapse of China itself. The nation faces severe challenges when it comes to food and energy security, both of which threaten to exacerbate a demographic death spiral. We have never witnessed an emerging superpower flame out overnight, though, nor have we been so economically tied to what would be a failed state. The collapse of the Soviet Union was a slow-motion train wreck, one which we were insulated from geographically, economically, and culturally. The sudden failure of China would represent an enormously dangerous moment in global history.

The best outcome is for America to rebuild our national strength, reassert our global leadership, and rein in Beijing's worst abuses. Doing so would allow a nation of 1.1 billion people to continue to grow and thrive, but not at the expense of our people. There is little ambiguity in how we achieve this—I've been talking and writing about it for years. Donald Trump made progress in reshaping the global narrative on China, and there is some bipartisan agreement for a pro-American industrial policy, but to be successful we need to focus on this challenge instead of whatever fake crisis the Marxist academics come up with next. A strong America is the world's greatest hope for a peaceful rise of China.

When I sat down to write this book, I began thinking about all the places where things went wrong in the past. Just like my grandfather said: Study history, learn from it, and act on that knowledge. I began this chapter with Ronald Reagan's famous "time for choosing" speech because I think it is important to remember—not just now, but always—that as new and dire as our circumstances might seem, they are almost never completely new. The nation has undergone radical change before, and we have always come out the other side stronger. The Republican Party has also been forced to rethink its identity in the past, always ending the process leaner and more resilient than when it began.

As Oren Cass pointed out at the beginning of 2022, the process of realignment within political parties is not unlike the way science—*real* science, not the MSNBC-approved platitudes—arrives at new discoveries that upend the old established order of things. Quoting research from the 1960s about scientific discoveries, Cass writes, "Science doesn't steadily evolve and improve; it lolls in long 'static' periods during which a community of researchers works mostly to validate their existing paradigm, punctuated by short periods of disruption when an old paradigm fails and a new one emerges."[28]

In other words, it is not uncommon for the "experts" to allow bad ideas to proliferate in any given field—even politics—defending them only because their careers and reputations depend on them. When that is no longer possible, a new paradigm emerges.

That, in a sense, is what is happening in today's Republican Party. Ideas that were once shunned for being "not Republican enough" are now widely accepted. Even the most staunch defenders of the old guard have acknowledged that the world is changing rapidly, and that those in government need to act quickly to address these changes.

I have been proud to lead the charge on many of these key issues, and I will continue to do so as I begin my third term in the US Senate.

As I see it, we have three main tasks ahead of us. First, we need to rebalance America's domestic economy by putting Wall Street in its place. Throughout this book, I have made the case that America's capital markets too often operate like a Ponzi scheme that extracts short-term profits while leaving our country, communities, and small businesses destitute. At their best, of course, markets are the lifeblood of our country. They facilitate growth and innovation by connecting people who have good ideas to those with money to invest in good ideas. But in recent years, Wall Street has perverted this mission by funneling cash into companies either directly or indirectly controlled by the Chinese Communist Party. No nation can survive by giving jobs and money to an adversary. It was naive when it began a few decades ago, and it is dangerously stupid now.

Our second task is to bring critical industries back to America. Some may be tempted to call this an overreaction to Covid-induced supply chain disruptions. But while the pandemic certainly caused people to open their eyes, the need for revitalizing domestic production has been obvious for years—albeit only to those willing to defy Washington's conventional wisdom. We need an aggressive pro-America industrial policy.

Finally, we have an obligation to rebuild America's workforce. For one, it is impossible to restore American industry without a strong, dynamic workforce. But work is also about human dignity. Globalization and outsourcing have destroyed the only reliable path to a stable and prosperous life available to millions of Americans. Families have broken down. Communities have crumbled, and death and despair have taken their place. We need to chart a new course by restoring the dignity of work.

Clearly, we face a lot of challenges. But despite them all, I am optimistic about America's future. There is no other nation in the world where I, the son of an immigrant bartender and stay-at-home mom, would be in this position. And I'll be damned if I am going to stand by while our nation becomes just a footnote in world history. Because the truth is that if America fails, I have nowhere left to go.

Fortunately, the process of looking back on history has given me reason for hope. It is not too late to correct course and sail away from the siren song of American Marxism.

But time is running out. America has been through tough times before, and we have always come out stronger on the other side. This time threatens to be different.

Never before have we faced enemies as dedicated to our collapse as we do right at this moment. It is not just a rising China or a belligerent Russia; radical woke Marxists are trying to destroy us from the inside. They are aided, abetted, and encouraged by our enemies, who know internal strife is the fastest way to ruin.

We must recommit ourselves to America's founding ideals, have pride in our great nation, and be willing to throw aside the stale policy ideas of the past to make this a new American century.

I have seen the look on the faces of people who lost their country, and I'll be damned if I will ever allow my children or their children to experience that shame and despair.

Acknowledgments

I thank my Lord, Jesus Christ, whose willingness to suffer and die for my sins will allow me to enjoy eternal life.

I have always believed that to live is to learn. This book is the result of my reflections of all the experiences I have had listening to and understanding the challenges of the many Americans I have come in touch with as a United States senator as well as a candidate for both the Senate and presidency of the United States. I would like to thank all the people whom I have met on these journeys and the talented people who have worked with me in all of these capacities.

I am grateful to Keith Urbahn and his team at Javelin for their help in the publication of this book. Eric Nelson is a tremendous editor, and his efforts greatly improved our manuscript. I am grateful to him, as well as to James Neidhardt and the entire team at HarperCollins, that worked on this book. Sean McGowan helped organize and craft the manuscript, and pull all of the ideas into a coherent narrative.

I do not believe any Senate office has been more principled, creative, or productive than my Senate office in confronting the challenges our nation faces today and coming up with common-sense solutions to meet them. I am grateful to Mike Needham, Dan Holler, Lauren Reamy, Caleb Orr, James Hitchcock, Collin Slowey, Brian Walsh, Will Green, Viv Bovo, Wes Brooks, Haim Engelman, Chris Griswold, Peter Mattis, Laura Ortiz, Bethany Poulos, Ryan Rasins, Ansley Rhyne, Samantha Roberts, Caleb Seibert, Connor Tomlinson, Jaime Varela and Meredith West, Brendan Hanrahan, Tim Shinbara, Phillip Todd, along with so many others, for their work.

An incredible team of individuals has helped me do outreach and constituent services in the State of Florida. I'm grateful for Todd Reid, Elena Crosby, and all the hardworking members of my Florida team. Jessica Fernandez, Chris Howd, Virginia Hinzman, and Bridget Spurlock have always gone the extra mile to keep our operations running smoothly.

One of the great privileges of being a United States Senator is the number of bright minds that are willing to spend time working with me and my office. I would like to thank Oren Cass, Ryan Anderson, Rob Atkinson, Dan Babich, Jonathan Baron, Jonathan Berry, Patrick Brown, Bridge Colby, Robert Doar, David Feith, John Garnaut, Robert C. Hockett, Alex Joske, Julius Krein, Yuval Levin, Ambassador Robert Lighthizer, Michael Lind, Dimon Liu, Stephen Miller, Henry Olsen, Gladden Pappin, Matthew Pottinger, David Rader, Rusty Reno, Reihan Salam, Randy Schriver, Michael Stumo, Nurel Turkel, Russ Vought, and Brad Wilcox for their collaboration.

My friend Norman Braman has always provided me wise counsel over the years.

Last, but certainly not least, my family—Jeanette, my wife; and our children, Amanda, Daniella, Anthony, and Dominick—have continued their love, support, and understanding.

NOTES

1. https://www.epi.org/publication/growing-china-trade-deficits-costs-us-jobs/.
2. https://www.economist.com/united-states/2022/04/09/black-americans-have-overtaken-white-victims-in-opioid-death-rates.
3. https://www.cfr.org/article/ten-lessons-return-history.
4. https://s.wsj.net/public/resources/documents/stanfordlanguage.pdf.
5. https://www.spglobal.com/spdji/en/documents/research/research-sp-examining-share-repurchases-and-the-sp-buyback-indices.pdf.
6. https://www.nytimes.com/2010/09/03/business/03commission.html.
7. Andrew Ross Sorkin, *Too Big to Fail: The Inside Story of How Wall Street and Washington Fought to Save the Financial System—and Themselves* (New York: Penguin), 2009, 5.
8. Peter Schweizer, *Red Handed: How American Elites Get Rich Helping China Win* (New York: HarperCollins), 2022, 140.
9. https://www.thenation.com/article/world/coronavirus-us-china-response/.
10. https://www.thefp.com/p/the-hijacking-of-pediatric-medicine?utm_source=direct&utm_campaign=post&utm_medium=web.
11. https://www.thefp.com/p/the-hijacking-of-pediatric-medicine.
12. https://opa.hhs.gov/sites/default/files/2022–03/gender-affirming-care-young-people-march-2022.pdf.
13. https://www.nationalreview.com/corner/major-swedish-hospital-bans-puberty-blocking-for-gender-dysphoria/.
14. https://uvahealth.com/services/transgender/transgender-ftm-surgery.
15. https://www.washingtonpost.com/news/to-your-health/wp/2018/02/28/trans gender-surgeries-are-on-the-rise-says-first-study-of-its-kind/.
16. https://www.nytimes.com/2022/06/10/science/transgender-teenagers-national-survey.html.
17. Ibid.
18. https://nypost.com/2022/06/18/detransitioned-teens-explain-why-they-regret-changing-genders/.
19. https://www.caranddriver.com/news/a31994388/us-auto-industry-medical-war-production-history/.
20. https://blogs.loc.gov/inside_adams/2016/04/when-a-quote-is-not-exactly-a-quote-general-motors/.
21. https://www.apple.com/customer-letter/.
22. https://nypost.com/2021/04/02/mlb-all-star-game-out-of-atlanta-over-georgias-voting-law/.
23. https://nypost.com/2021/04/12/top-corporate-execs-met-on-zoom-to-mount-new-voting-legislation-effort/.
24. https://www.washingtonexaminer.com/news/coca-cola-training-be-less-white.

25. https://www.penguin.co.uk/articles/2020/06/ibram-x-kendi-definition-of-antiracist.
26. https://www.nytimes.com/2020/02/20/opinion/taliban-afghanistan-war-haqqani.html.
27. https://www.bariweiss.com/resignation-letter.
28. https://americancompass.org/2022-founders-letter/.

Index

abortion, 27, 60, 141, 157, 163
AFL-CIO, 149
Alliance for Prosperity, 125
Alvarez, Max, 113
Amazon, 82, 91, 92, 150
American Academy of Pediatrics (AAP), 101
American Compass, 18
American Dream, 1–2, 8, 12, 108, 155, 196–197
American Enterprise Institute, 45
American Investment in the 21st Century, 28
American Marxism, 170–173, 180–181, 190–192, 206, 208
American Revolution, 9–10, 187
American System, 11, 15
amnesty programs, 116–118, 119, 122–123, 149
Anderson, Ryan T., 150
anticommunism, 64
Apple, 51, 138–140
assimilation, 118–119, 128
Atlantic, The, 179

Baker, James, 52–53
Balkan Wars, 68
banking industry, 27–35
  and 2008 financial crisis, 35–43
Batista, Fulgencio, 110
Bell, Derrick, 182, 185
Bennet, James, 190
Bernanke, Ben, 36
Bernstein, Carl, 187
Best and the Brightest, The (Halberstam), 63–64
Between the World and Me (Coates), 179
Biden, Hunter, 49, 74–75
Biden, Joe, 78, 96, 174
  2020 presidential campaign, 123, 192
  cabinet nominations, 16
  economic policies of, xx
  on PPP, 93
  presidency, 202–205
  support for China by, 15
  vice presidency, 162–163
Biden administration, 59, 78, 141, 149
  Chinese tariffs, 17
  immigration policies, 122–127, 133–134
  on social justice issues, 105–106
  support for gender-affirming care, 101–105
Bishop, Bill, 88
Black Lives Matter movement, 152–153, 178, 180
BlackRock, 144, 153
Blinken, Anthony, xxi
border security, 121–123
Bouie, Jamelle, 160
Brady, Anne-Marie, 50
Bridgewater Capital, 50
Brown, Michael, 176–177
Brownstein, Ronald, 163
Brunswick Group, 154
Buchanan, Patrick, 56–57
Bush, George H. W., 52, 54–55, 59, 65, 68
Bush, George W., 40, 58, 68, 74
Bush administration (1989–1993), 53, 54

Bush administration (2001–2009), 37, 39, 156
Business Insider, 123
Business Roundtable, 17, 143–144

Cameron, David, 62
campaign financing, 24, 26–27
Campbell, Kurt, 59
capitalism, 29, 53
  common-good, 43–47, 147, 197
  free market, 147–148
  Smith's theory of, 18
  stakeholder, 141–148, 153
Carter, Jimmy, 193, 195
Cass, Oren, 4, 18, 146, 207
Castro, Fidel, xiv–xv, 110–114, 190
Catholic Charities of Miami, 112–113
Catholic University of America, 147
CCP. *see* Chinese Communist Party
CEFC China Energy, 49
Cena, John, 62
CEO compensation, 34
Chamber of Commerce, 17
Child Tax Credit, 145, 147, 197–198
China, xxi, 39
  Apple and, 138, 140
  Belt and Road Initiative, xxii, 81
  Chinese companies and the stock exchange, 198
  Covid-19 pandemic in, 88–90, 98
  dependence on rare earth minerals from, 47, 82, 83
  espionage campaigns, 78–79
  global ambitions of, 52–63, 72–74, 81, 85
  human rights concerns, xix, 56
  power conflict between America and, 72–74
  rise of at our expense, 48–51, 61–66, 69
  Russia's partnership with, 71
  and Taiwan, 72, 85–8762, 204
  tariffs on, 17
  threats from, 78–84
  Tiananmen Square protests, xvi–xvii, 5, 53–55
  Trump's trade deal with, 79–80
  Uyghur internment camps in, 20–21, 60, 80–81
  WTO entry, 2, 5–6, 9, 13–15, 18, 48–49, 116
China Initiative, 78
Chinese Communist Party (CCP), xvi–xvii, 2, 20–21, 49–50, 52, 54–55, 57–63, 71, 85–87, 153, 206–207
Citibank, 72
Citigroup, 34
Civil Rights Act (1964), 169
Clapper, James, 75, 77
Claremont McKenna College, 127–131
Clay, Henry, 10–11, 15
Clinton, Bill, 6, 34, 56–57, 65
Clinton, Hillary, 43, 44, 70, 76, 77, 120, 159–160, 201–202
Clinton administration, 33, 57–58, 156
CNN, 75, 189
coalition of the ascendant, 164, 165
Coates, Ta-Nehisi, 179
Coca-Cola, 51, 80, 152, 154
Coddling of the American Mind, The (Haidt & Lukianoff), 130
Codina, Armando, 113
Cold War, xi–xii, xiv, 67, 84, 165
colleges and universities
  identity politics in, 128–131
  Marxism in, 181–186
Columbia University, 129
  Columbia Business School, 8
common-good capitalism, 43–47, 147, 197
communism, xvi–xviii, 49–50,

52, 68, 111–114, 180. *see also* Chinese Communist Party
community life, 22
Congressional Record, 137
Conservative, Inc., 195–199
consumerism, 16–17
Cook, Tim, 139–140
corporate governance, 141–148
corporate profits, 17–19, 27–28, 32, 41, 56, 80–81, 137–138
"Cost of Thriving Index," 4
Cotton, Tom, 17, 189–191
Covid-19 pandemic, 5, 60, 78–79, 88–100, 141, 192, 203, 208
Crenshaw, Kimberlé, 185
critical race theory, 31, 111, 152, 179, 182–183, 185, 189
Cuba, xiv–xv, 110–114, 132–134
Cuban American National Foundation, xv
Cuban Democracy Act, xv
Cuban Miracle, 31

Dalai Lama, 62
Dalio, Ray, 50–51
decadence, xiv
Declaration of Independence, 114
Delta Airlines, 142
Deng Xiaoping, 52, 54, 60
Derrida, Jacques, 182
DiAngelo, Robin, 152, 160
Dimon, Jamie, 144, 152–153
Disney, 60, 80
Dobbs v. Jackson Women's Health Organization, 141
Donne, John, xii
donor class, 27–29
Douthat, Ross, xiv
DREAM Act, 117

East Germany, xvi
economic policy
Biden's, xx
Clay's vision, 11
for a healthy economy, 43
neoliberal, 13
in support of a free market, 145–148
viewing Americans as primarily consumers, 16–17
Economic Policy Institute, 14
economy
2008 financial crisis and, 35–43
financialization of, 29–35, 37, 40, 42
PPP benefits to, 91–94
signs of healthy, 42–43
education, 111–112
Eisenhower, Dwight D., 135
elections
1980 presidential, 193–194, 196
2000 presidential, 68–69
2008 presidential, 156–159
2012 presidential, 42, 163, 165
2016 presidential, 2, 8–9, 44, 75–77, 119, 145, 159–160, 180, 199–202
2018 midterm, 164
2020 presidential, 74–75, 123, 164
financing of, 24, 26–27
party decides model, 23–24
elites
as the donor class, 27–29
and the "end of history" conflict, xiii–xiv
othering of working class by, 157–160
Ellis Island, 113, 114
Engels, Max, 181
environmental, social, and governance (ESG) initiatives, 144, 146–147
equity, 172–173
ethnonationalism, 118

experts, 94–95, 100, 106–107, 108, 208. *see also* technocrats
ExxonMobil, 144

Facebook, 75, 77, 140
Fannie Mae, 39
Farook, Syed Rizwan, 139
Fauci, Anthony, 88, 91, 97, 100, 101–103, 109
FBI, 139
Federal Reserve, 39, 92, 95
fentanyl, xx, 121, 204, 206
Ferguson, MO protests, 176–177
finance industry, 27–35
  and 2008 financial crisis, 35–43
financial crisis (2008), 25, 27–28, 35–43, 95
Fink, Larry, 153
Florida A&M University, 7
Floyd, George, xxi, 180, 189
Ford Motor Company, 32
Foreign Affairs, 59
foreign policy
  America's, 68–74
  China's, 60
  neoliberal, 13
Foroohar, Rana, 33–34
Fortune, 9
Fortune 500, 143
Freddie Mac, 39
free market, 13, 15, 39, 45, 48, 53, 64, 93, 108, 145–148
Free Press, 101
Friedman, Thomas, xviii, 19–20, 53, 116
Fukuyama, Francis, xi–xii, xvii, 13

Gallup, 161
Gang of Eight bill, 121–122
gay marriage, 162, 163, 167–171
Geithner, Timothy, 40
gender-affirming care, 100–107
General Motors, 135–137
genocide, xix, 68, 73, 80, 153
George Floyd protests, 180, 189
globalization, xx, 14–15, 33, 39, 115, 143, 208
Goldman Sachs, 40, 61
Google, 9
Gorbachev, Mikhail, xii
Gore, Al, 68
Gorsky, Alex, 144
Great Depression, 30, 37
Guevara, Che, 111, 190
Gulf War, xviii, 67–68
Gutierrez, Luis, 117–118, 131

Haidt, Jonathan, 130
Halberstam, David, 63–64
Hamilton, Alexander, 9–10
Hannah-Jones, Nicole, 186–187
Haqqani, Sirajuddin, 191
Harris, Kamala, 204–205
Harvard Business School, 32
Harvard Law School, 181
Hass, Richard, 19
Hawley, Josh, 17
Hegel, Georg Wilhelm Friedrich, xi
Helms, Jesse, 15
Hillbilly Elegy phenomenon, 6
Hitler, Adolf, 68
Holder, Eric, 169
Holtz-Eakin, Doug, 92
home ownership, 37–38. *see also* financial crisis (2008)
House Permanent Select Committee on Intelligence, 76–77
How to Be an Antiracist (Kendi), 172
Hu Jintao, 58
Hussein, Saddam, xviii, 67

ICE. *see* U.S. Immigration and Customs Enforcement
identity politics, 128–131, 153, 163
immigration
  asylum process, 124–125

and coalition of the ascendant, 164, 165
from Cuba, 110–114, 132–134
illegal, 115–119, 204–205
patriotism of immigrants, 132–134
reform, 119–127
Title 42, 126
Industry Week Magazine, 12
Instagram, 105, 140
International Review of Psychiatry, 105
international trade, 5–6, 13–15, 58
intersectionality, 185
Iran, xix–xx
Irreversible Damage (Shrier), 106
It's a Wonderful Life, 30
ivermectin, 102, 103
James, LeBron, 62
Jiang Zemin, 58
Johnson, Lyndon, 64
Johnson & Johnson, 144
J.P. Morgan, 144

Kaepernick, Colin, 80
Kendi, Ibram X., 172–173
Kennedy, Anthony, 169–170
Kennedy, John F., 64
Kennedy, Ted, 193–194
King, Martin Luther, Jr., 128, 176–177, 179, 188
Kotkin, Joel, 64
Krauthammer, Charles, xviii
Krein, Julius, 146
Kristol, Irving, 45

labor supply, 115–119
labor unions, 149–152
Ladapo, Joseph, 106
Lavrov, Sergey, 70
League of Nations, 84
Leeser, Oscar, 126
Lenin, Vladimir, 49
Leo XIII, Pope, 147
Levi Strauss, 142
Library of Congress, 137
Lind, Michael, 10
LinkedIn, 142
Lukianoff, Peter, 130
Lynch, Loretta, 77

Ma, Jack, 50, 60
MacArthur Genius Grant, 179
"Made in China 2025 and the Future of American Industry," 21
Major League Baseball (MLB), 142
Makers and Takers (Foroohar), 33
Malik, Tashfeen, 139
manufacturing industry
effects of China's WTO entry on, 5–9, 14–15, 18
in Hialeah, FL, 3–4, 11–13
job shrinkage in, 32–33
offshoring of, 13, 25, 115
outsourcing of, 13, 47
post-war, 11
PPP benefits to, 93
in Revolutionary America, 9–10
Mao Zedong, 60, 63
Marx, Karl, 134, 181, 183, 184
Marxism
American, 171–173, 180–181, 190–192, 206, 208
as a power model, 173–175, 181–182
in universities, 181–186
McCain, John, 160
McDonald's, 19–20, 55, 72
McKinsey & Company, 60
Medium, 7
Merkley, Jeff, 21
Meta, 9, 140–141
Miami Freedom Tower, 113
Mian, Atif, 37
microaggressions, 129

Morey, Daryl, 62
mortgage-backed securities, 35–37
MSNBC, 75, 202, 207
Mulan, 60, 80

Nation, The, 98
National Affairs, 139
National Football League
 commissioner, 46
National Institutes of Health
 (NIH), 96, 102
National Interest, xi, xv
National Journal, 163
National Public Radio, 75
national security, 82–84
NATO, 68
neoliberalism, xiii–xiv, 2, 13, 55, 64,
 70
New Deal, 94
Newsom, Gavin, 99
Newsweek, 117
New York Post, 74, 105, 142, 154
New York Times, xiv, xviii, 19, 93,
 104, 116, 137, 160, 183, 188–
 191, 193, 202
New York Times Magazine, 186–187
Nike, 51, 72, 80, 138
Nixon, Richard, 52, 187
North American Free Trade
 Agreement (NAFTA), 13, 116

Obama, Barack, 24, 34, 59, 75, 177
 2008 presidential campaign,
 156–161
 2012 presidential campaign,
 70–71, 162–165
 presidency, 39–40, 117, 119,
 121–122, 161–162, 165–167,
 174, 175, 179–180
Obama administration, 76, 167–171,
 199
Obergefell v. Hodges, 169–171
Ocasio-Cortez, Alexandria, 120
offshoring, 13, 25, 115, 154
open borders policy, 116–117, 131,
 204–205
Operation Peter Pan, 112–113, 114
opioid epidemic, xx, 8
organized labor, 149–152
outsourcing, 13, 47, 153–154, 208

paper wealth, 36–37
patriotism, xiv, 81, 112, 115, 126,
 132, 138, 143, 155
Paulson, Hank, 40
Paycheck Protection Program
 (PPP), 91–94
Pelosi, Nancy, 99
Pentagon Papers, 187
Pew, 164
Pocan, Mark, 120
police shootings, 176–179
PPP. *see* Paycheck Protection
 Program
prosperity gospel, xvii–xviii
Psaki, Jen, 131
Psalms, Book of, 159
puberty blockers, 101–103, 106
Pulitzer Prize, 187
Putin, Vladimir, xvi, xvii, xix, 20, 49,
 70–72, 76–77, 83, 204

Racial Equity and Justice Initiative,
 140
racial wealth gap, 7
railway labor dispute, 149–150
rare earth minerals, 47, 82, 83
Ratner, Ely, 59
Reagan, Ronald, 65, 193–196,
 201–202, 205–206, 207
Red Handed (Schweizer), 50
Republican National Committee,
 165–166
Reuters, 98
Romney, Mitt, 42, 70, 71, 76, 165
Roosevelt, Franklin Delano, 94

Rubin, Robert, 33–34
ruling class, 24–25
Russia, xvi, xvii, xix, xxi, 39, 70–71, 73, 76
invasion of Ukraine, xix, 20, 70–72, 83, 204
propaganda efforts by, 75–77

safe spaces, 184
Sanders, Bernie, 15, 17, 28, 43, 111, 119–120, 134
SBA. *see* Small Business Administration
Scalia, Antonin, 169
Schiff, Adam, 76–77
schools, 111–112, 203
Schwartz, Yishai, 139–140
Schweizer, Peter, 50
science, 96–100, 103, 106–107, 203, 207
Scowcroft, Brent, 54
Senate Armed Services Committee, 136
Senate Small Business and Entrepreneurship Committee, 89–91, 92, 146
September 11 attacks, 69, 74
Sheehan, Neil, 187
Shrier, Abigail, 106
1619 Project, 186–188
Slate, 160
slavery, 187–188
Small Business Administration (SBA), 89–91
Smith, Adam, 18
socialism, 44–45, 134, 173, 175
social media, 75, 168, 177–178, 181
Sonnenfeld, Jeffrey, 142, 143
Sorkin, Andrew Ross, 36, 40
Soviet Union, xii, xv–xvii, 53, 65, 67, 70, 183, 206–207
Spellman, Mary, 129–130
stakeholder capitalism, 141–148, 153
Stalin, Joseph, 198
Stanford University, Elimination of Harmful Language Initiative, 25
stock market, 17–18, 25, 32–34, 39, 144–145
successor ideology, 185–186
Sufi, Amir, 37
Summers, Lawrence, 33–34
supply chain disruptions, 17, 89–90, 208
systemic racism, 151–155, 163, 178

Tabak, Lawrence, 102–103
Taiwan, 62, 72, 85–87, 204
Target, 106
Tax Cuts and Jobs Act (2017), 145
taxes, 33–34, 46–47
Child Tax Credit, 145, 147, 197–198
corporate, 154, 195, 197
reform, 197
tech industry, 9
technocrats, 94–100, 107
terrorism, 69, 73, 74, 139
Tesla, 51, 82
Thomas-Greenfield, Linda, xxi
Tiananmen Square protests, xvii, 5, 53–55
Tibet Autonomous Region, 58, 62
Title 42, 126
Torricelli, Bob, xv
transgender individuals
application of Civil Rights Act (1964) to, 169
and gender-affirming care, 100–107
students, 203
trigger warnings, 184
Trump, Donald, 43, 76–78, 120, 160, 189, 192, 207

Trump, Donald (*continued*)
2016 presidential campaign, 8–9, 199–202
2020 presidential campaign, 164
immigration policies, 121–123
Trump administration, 17, 78, 79–80, 91
trust, erosion of, 74–77, 94, 97–98
Tumblr, 105
Twitter, 75, 77, 190
Two Cheers for Capitalism (Kristol), 45

Uber, 9
Ukraine, Russian invasion of, xix, 20, 70–72, 83, 204
unemployment, 17, 40, 91–92
unipolar moment, xviii–xix
United Nations, 68
University of California, 129
U.S. Chamber of Commerce, 116
U.S. Customs and Border Patrol, 122–123, 125
U.S. Department of Defense, 135–136
U.S. Department of Education, 96
U.S. Department of Health and Human Services, 96
U.S. Department of Homeland Security, 96
U.S. Department of Justice, 78
U.S. Department of State, 52–53
U.S. Environmental Protection Agency, 96
U.S. Food and Drug Administration, 103
U.S. Immigration and Customs Enforcement (ICE), 120–121
U.S. Supreme Court, 141, 169–171
Uyghur Forced Labor Prevention Act (2018), 21

values, xii–xiv, 22, 24, 53, 81, 112, 114, 121, 126, 148, 152–153, 159–160, 171–173, 201
Vance, J. D., 6
Vietnam War, 63–64, 187
voting laws, 141–142, 154

wage stagnation, 32–33, 37, 41
Wall Street, 31, 35
Wall Street Journal, 116, 197
Walmart, 142
Warner Bros., 153
Warren, Elizabeth, 120–121
Washington, George, 10
Washington Post, 93, 104, 146
Wealth of Nations, The (Smith), 18
Weiss, Bari, 189, 191
Westinghouse Electric Company, 135
When Harry Became Sally (Anderson), 150
white fragility, 160–161
White Fragility (DiAngelo), 152, 160
White House Coronavirus Task Force, 88, 100
Wilson, Charles E., 135–137
Wilson, Woodrow, 84
wokeness, 185
corporate, 78–81, 141–148, 151, 152–155
in the military, 204
Woodward, Bob, 187
workforce
immigrant labor supply, 115–119
labor unions and, 149–152
rebuilding of, 148–149, 208–209
working class
issues important to, 25–26
othering of by elites, 157–160
rise and fall in prosperity of, 30–33
voters, 156–157

World Trade Organization (WTO), 2, 5–6, 9, 13–15, 18, 48–49, 116
World Transformed, A (Bush), 54
World War I, 84, 135
World War II, 11, 67, 68, 135–136, 139
WTO. *see* World Trade Organization

Xi Jinping, 20–21, 49, 59–63, 79–81, 98, 140

Yale University, 54, 58
School of Management, 142
Yang, Wesley, 185–186
Ye Jianming, 49

Zoellick, Robert, 52–53, 58–59, 61
Zuckerberg, Mark, 75, 140

About the Author

Marco Rubio has represented Florida in the United States Senate since 2010, where he has one guiding objective: bring the American Dream back into the reach of those who feel it slipping away.

Senator Rubio's efforts have been successful and long-lasting. Nonpartisan analyses by GovTrack and the Center for Effective Lawmaking have ranked Rubio the Senate's number two leader and most effective Republican.

Senator Rubio currently serves as vice chairman of the Senate Select Committee on Intelligence, where he oversees our nation's intelligence and national security apparatus. Senator Rubio is also a member of the Foreign Relations Committee, where he fights to promote human rights and America's interests around the globe; the powerful Appropriations Committee, which allocates funding for the federal government; and the Special Committee on Aging, dedicated to the needs of older Americans.

In addition, Senator Rubio serves on the Committee on Small Business and Entrepreneurship, which works to help small businesses thrive in the twenty-first century. As the former chairman of this committee, Rubio authored the historic Paycheck Protection Program, which has been a lifeline to millions of small businesses and American workers as they battle economic hardship in the wake of the Covid-19 pandemic.

Senator Rubio was born in Miami, after his parents came to the United States from Cuba in search of the American Dream. He lives there today with his wife, Jeanette, and their four children.